T0238360

Springer Series in
ADVANCED MICROELECTRONICS 22

Springer Series in
ADVANCED MICROELECTRONICS

Series Editors: K. Itoh   T. Lee   T. Sakurai   W.M.C. Sansen   D. Schmitt-Landsiedel

The Springer Series in Advanced Microelectronics provides systematic information on all the topics relevant for the design, processing, and manufacturing of microelectronic devices. The books, each prepared by leading researchers or engineers in their fields, cover the basic and advanced aspects of topics such as wafer processing, materials, device design, device technologies, circuit design, VLSI implementation, and subsystem technology. The series forms a bridge between physics and engineering and the volumes will appeal to practicing engineers as well as research scientists.

18 **Microcontrollers in Practice**
    By I. Susnea and M. Mitescu

19 **Gettering Defects in Semiconductors**
    By V.A. Perevoschikov and V.D. Skoupov

20 **Low Power VCO Design in CMOS**
    By M. Tiebout

21 **Continuous-Time Sigma-Delta A/D Conversion**
    Fundamentals, Performance Limits and Robust Implementations
    By M. Ortmanns and F. Gerfers

22 **Detection and Signal Processing**
    Technical Realization
    By W.J. Witteman

Volumes 1–17 are listed at the end of the book.

W.J. Witteman

Detection and Signal Processing

Technical Realization

With 63 Figures

 Springer

Professor Dr. Wilhelmus Jacobus Witteman
Universiteit Twente
Postbus 217
7500 AE Enschede
The Netherlands
E-mail: W.J.Witteman@tnw.utwente.nl

Series Editors:

Dr. Kiyoo Itoh
Hitachi Ltd., Central Research Laboratory, 1-280 Higashi-Koigakubo
Kokubunji-shi, Tokyo 185-8601, Japan

Professor Thomas Lee
Stanford University, Department of Electrical Engineering, 420 Via Palou Mall, CIS-205
Stanford, CA 94305-4070, USA

Professor Takayasu Sakurai
Center for Collaborative Research, University of Tokyo, 7-22-1 Roppongi
Minato-ku, Tokyo 106-8558, Japan

Professor Willy M. C. Sansen
Katholieke Universiteit Leuven, ESAT-MICAS, Kasteelpark Arenberg 10
3001 Leuven, Belgium

Professor Doris Schmitt-Landsiedel
Technische Universität München, Lehrstuhl für Technische Elektronik
Theresienstrasse 90, Gebäude N3, 80290 München, Germany

ISSN 1437-0387

ISBN 978-3-642-06737-2   e-ISBN 978-3-540-29600-3

This work is subject to copyright. All rights are reserved, whether the whole or part of the material is concerned, specifically the rights of translation, reprinting, reuse of illustrations, recitation, broadcasting, reproduction on microfilm or in any other way, and storage in data banks. Duplication of this publication or parts thereof is permitted only under the provisions of the German Copyright Law of September 9, 1965, in its current version, and permission for use must always be obtained from Springer-Verlag. Violations are liable to prosecution under the German Copyright Law.

Springer is a part of Springer Science+Business Media.

springeronline.com

© Springer Berlin Heidelberg 2006
Softcover reprint of the hardcover 1st edition 2006

The use of general descriptive names, registered names, trademarks, etc. in this publication does not imply, even in the absence of a specific statement, that such names are exempt from the relevant protective laws and regulations and therefore free for general use.

Cover concept by eStudio Calmar Steinen using a background picture from Photo Studio "SONO".
Courtesy of Mr. Yukio Sono, 3-18-4 Uchi-Kanda, Chiyoda-ku, Tokyo
Cover design: *design & production* GmbH, Heidelberg

To my wife Nel

Preface

The writing of this book has been inspired by the experience of teaching a course on *Detection and Signal Processing* to graduate students over a period of many years. It was striking that students were not only fascinated by the various detection principles and technical performances of practical systems, but also by the professionalism of the involved typical physical engineering. Usually students are thoroughly taught in different courses of physics, which are mostly studied as isolated fields. The course on detection and signal processing is based on typical results that were established in different disciplines like optics, solid state physics, thermodynamics, mathematical statistics, Fourier transforms, and electronic circuitry. Their simultaneous and interdependent application broadens the insight of mutual relations in the various fields. For instance the fluctuations of thermal background radiation can be derived either with the black body theory or independently with thermodynamics to arrive at the same result. Also the applied Fourier relations in the frequency and time domains are no longer abstract mathematical manipulations but practical tools and probably easier to understand in the applied technique. Estimates of the order of magnitudes with comparison of relevant physical effects necessary by designing a device are very instructive. In general the achievements of various disciplines are brought together to design and to evaluate quantitatively the technical performances of detection techniques. Thus the interest for detection and signal processing is both to learn the knowledge for designing practical detection systems and to get acquainted with the thinking of physical engineering.

The first part of the book is devoted to noise phenomena and radiation detectors. Fundamental descriptions with quantitative analyses of the underlying physical processes of both detectors and accompanying noise lead to understand the potentials with respect to sensitivity and operating frequency domain. The second part deals with amplification problems and the recovery of repetitive signals buried in noise. The last part is devoted to solving the problems connected with reaching the ultimate detection limit or quantum limit. This is done for heterodyne detection and photon counting. Although heterodyne detection yields the ultimate sensitivity, its spatial mode selectivity and, in general, the low spectral power density of the signal require sophisticated provisions. This is discussed in detail. The inherent problems are analyzed and appropriate technical solutions are described to reach the ultimate sensitivity for detecting incoherent radiation and communication signals that are randomly Doppler shifted. The results are illustrated with examples of space communication.

Hengelo (O), January 2006 *W. J. Witteman*

Contents

1 **Random Fluctuations** 1
   1.1 Introduction 1
   1.2 Thermal Noise of Resistance 2
   1.3 Shot Noise 5
        1.3.1 Spectral Distribution 6
        1.3.2 Photons 10
   1.4 Flicker Noise 10
   1.5 Generation–Recombination Noise 10
   1.6 Thermal Radiation and Its Fluctuations 13
   1.7 Temperature Fluctuations of Small Bodies 18
        1.7.1 Absorption and Emission Fluctuations 20

2 **Signal–Noise Relations** 21
   2.1 Signal Limitation 22
   2.2 Background Limitation 22
        2.2.1 Ideal Detection 24
   2.3 Johnson Noise 27
   2.4 Dark Current Noise 27
   2.5 Noise and Sensitivity 28
   2.6 Amplifier Noise and Mismatching 28

3 **Thermal Detectors** 31
   3.1 Thermocouple and Thermopile 31
   3.2 Bolometer 36
        3.2.1 Metallic Bolometer 39
        3.2.2 Thermistor 40
   3.3 Pyroelectric Detector 44

4 **Vacuum Photodetectors** 51
   4.1 Vacuum Photodiode 52
   4.2 Photomultiplier 56

5   Semiconductor Photodetectors 61
   5.1   Photoconductors 61
      5.1.1   Analysis of the Detection Process.................... 64
      5.1.2   Frequency Response 69
   5.2   Photodiodes 69
      5.2.1   P–N Junction 70
      5.2.2   Current–Voltage Characteristic..................... 72
      5.2.3   Photon Excitation 75
      5.2.4   Operational Modes 79
      5.2.5   Open Circuit 80
      5.2.6   Current Circuit 82
      5.2.7   Reverse-Biased Circuit 83
   5.3   Avalanche Photodiodes................................. 86
      5.3.1   Multiplication Process 87
      5.3.2   Multiplication Noise 89
      5.3.3   Detectivity 90
      5.3.4   Frequency Response 92

6   Correlation Analyses 95
   6.1   AutoCorrelation 95
   6.2   Cross Correlation 97
      6.2.1   Signal Recovery by Cross Correlation 99
      6.2.2   Periodic Signal Recovering by Autocorrelation 101
      6.2.3   Autocorrelation of White Noise 103
      6.2.4   Spectral Power Density from Shot Noise Correlation ... 104
      6.2.5   Correlations of Linear Detector Systems.............. 105

7   Signal Processing 107
   7.1   Operational Amplifier 108
   7.2   Lock-in Amplifier...................................... 111
      7.2.1   Two-Phase Lock-in Amplifier 115
   7.3   Signal Averagers 115
      7.3.1   Pulse Train Averagers 116
      7.3.2   Waveform Analyzer............................... 118
   7.4   Correlation Computer.................................. 119

8   Heterodyne Detection 121
   8.1   Analysis of Signal Conversion and Noise 122
   8.2   Signal Beam Profile.................................... 124
   8.3   Optical System.. 127
   8.4   Coherent versus Incoherent Detection 129
      8.4.1   Photodetectors.................................... 129
      8.4.2   Thermal Detector 130
      8.4.3   Pyroelectric Detector 131
      8.4.4   Heterodyne Detection of Incoherent Radiation 131

8.5   Heterodyne Lock-In Amplification 132
       8.5.1   High-Spectral Resolution 137
8.6   Dual Signal Beam Heterodyne Detection 138
8.7   Dual Signal Heterodyne Lock-In Amplification 145
8.8   Dual Signal Wave Analyzer 147
       8.8.1   Space Communication 148
       8.8.2   Transmitting Photographs 149
       8.8.3   Laser Radar 149

9   Fast Detection of Weak and Noisy Signals 153
    9.1   Suppressing Amplifier Noise with Detection Discriminator 154
    9.2   Photon Counting 156

A   Appendix 161
    A.1   Microcurrent Pulse 161
    A.2   Statistics 162
          A.2.1   Binomial Distribution 162
          A.2.2   Poisson Distribution 163
          A.2.3   Gaussian Distribution 164
          A.2.4   Photoelectron Statistics 165
    A.3   Multiplication Factor M_n 166
    A.4   Power Flow of Standing Wave Modes 167

References 169

Index 171

1

Random Fluctuations

1.1 Introduction

The sensitivity and accuracy of any detection system is limited by random fluctuations that always accompany the measurement. It also sets a limit to the minimum detectable signal. These random fluctuations or disturbing signals, called noise, can be divided into two categories depending on their nature. A part is not inherently connected to the detection principle but to the environment. Instrumental imperfections, atmospheric turbulence, vibrating mechanical constructions, 50 or 60 Hz and higher harmonics from the power line, radio and television stations, building vibrations, and temperature fluctuations all fall in this category. These environmental disturbances are in most cases occasional, peculiar to the surrounding, and not statistical. They can in principle be reduced to arbitrarily small values, but in practice, they can be very annoying and difficult to eliminate entirely. Reductions are often obtained by shielding or instrumental improvements.

The other category is fundamental from nature and inherently connected to the physical process that underlies the detection. For instance, through any conductor there is always a small fluctuating current due to the random thermal motion of the free electrons of this conductor. Other typical fluctuations arise from the fact that electrical currents are built up of irreducible elementary units, the charge of an electron. Similar effects occur for radiation as a flow of photons with discrete values. For this reason also thermal background radiation contains fluctuations and the temperature of a body is essentially not constant. Even in systems that filter out electronically the contributions of thermal background, a part of their fluctuations, are still present and mixed with the signal. The amount of this noise depends on fundamental physical quantities and sets the ultimate limit to the minimum detectable signal, which cannot be surpassed. Modern measuring instruments work close to their ultimate limits. Furthermore to exploit the sensitivity of a detection system, we must also ascertain the fundamental nature of the applied physical processes on which the detection is based.

For this reason we discuss in this chapter the fundamental aspects of noise. It will be done for thermal background radiation and for noise connected with the elementary units of radiation and charge carriers of the detection circuit. In detector circuits the electrons are not only driven by the incident radiation but also by random thermal motion. Further, the circuit currents from the signal of various detection systems have fluctuations due to the discreteness of the charge carriers and their random time distribution. Photoconductors produce additional noise by the random thermal process of generating carriers. Although various types of noise are fundamentally present in any detector system the interesting question is how can these contributions be minimized. For this purpose the physical nature of various noise sources will be treated in a qualitative and quantitative way. It will be done for background radiation, thermal electron motion present in any conductor, current fluctuations due to the discreteness of the electron charge and random photon emission in diodes, and for the generation and recombination processes in photoconductors. The derivation of the spectrum density and of the mean square fluctuations of the noise current turns out to be most relevant to detection systems. The total noise power still present in the final signal power of a detector system is then proportional to the frequency bandwidth of the system. The ratio of the output signal power to this noise power will be considered as the quality factor for the detection.

1.2 Thermal Noise of Resistance

Due to random motion of the electrons there are always fluctuations of the local charge density in any element of an electronic circuit. These charge densities cause voltage gradients which drive on their turn fluctuating currents. The average values of these fluctuations over large periods are, of course, zero, but this is not the case for a limited period. Intuitively one may say the smaller this time period or the larger the frequency bandwidth of the observation, the larger the fluctuations. The smallest period or upper frequency limit of this increasing thermal noise is set by the electron collision frequency which is roughly 10^{13} Hz. The thermal noise of conductors is the so-called Johnson noise. A quantitative treatment of this thermal noise can be carried out in different ways [1–3]. It is found that the thermal noise *power* of a resistive element with real impedance does not depend on resistivity, material, dimensions, or its surrounding but solely on its temperature and the frequency domain of the observation. The derivation is as follows.

Consider a closed loop containing a transmission line of length l connecting on both sides two identical resistors with resistivity R as illustrated in Fig. 1.1. The random thermal fluctuations of the electrons in the resistors can support traveling voltage waves in this closed loop. By choosing the characteristic impedance R_0 of the transmission line equal to R there are no reflections of waves at the ends. The natural frequencies of the loop correspond to

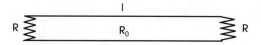

Fig. 1.1. Closed loop of transmission line

waves – called frequency modes – that are periodic in the round trip distance. Thus the wavelength is given by

$$\lambda_n = \frac{2l}{n}, \tag{1.1}$$

where n is an integer. The natural frequencies are then

$$\nu_n = \frac{nc}{2l}. \tag{1.2}$$

So the frequency spacing between these frequencies is $c/2l$. The number of traveling waves within a frequency bandwidth $\Delta\nu$ is then

$$N = \frac{2l\Delta\nu}{c}. \tag{1.3}$$

At thermal equilibrium the average energy of a single frequency mode is according to Planck's law

$$E_{h\nu} = \frac{h\nu}{e^{h\nu/kT} - 1}. \tag{1.4}$$

The total energy E_t in the bandwidth $\Delta\nu$ becomes $NE_{h\nu}$ or

$$E_t = \frac{2lh\nu\Delta\nu}{c(e^{h\nu/kT} - 1)}. \tag{1.5}$$

This energy moves with the velocity c so that the round trip time is $2l/c$. The power P flowing in each direction of the transmission line is then

$$P = \frac{h\nu\Delta\nu}{e^{h\nu/kT} - 1} \tag{1.6}$$

Usually the frequency of interest is much smaller than kT/h so that the spectral power density equal to $dP/d\nu$ can be considered constant and the noise is therefore often called "white noise" and is given by

$$P = kT\Delta\nu. \tag{1.7}$$

Since there are no reflections at the ends of the line the incoming power of a resistor is dissipated. Then an equal amount of power must also be generated by a resistor in order to have a balance of power. This power is apparently the thermal noise power of a resistor.

Let us now describe the resistor with its noise power by its resistance R in series with a noise generator having a mean square voltage amplitude $\overline{v_n^2}$.

The noise power P of the resistor that is delivered to a transmission line with an arbitrary characteristic impedance R_0 is then given by

$$P = \frac{\overline{v_n^2} R_0}{(R + R_0)^2} \cdot \tag{1.8}$$

It is seen that the maximum value of P is obtained for $R = R_0$, so that using (1.7) we find for the mean square voltage amplitude of a resistor

$$\overline{v_n^2} = 4kTR\Delta\nu . \tag{1.9}$$

An equivalent circuit of a resistor can be given by a noise current generator in parallel with the resistor. The mean square noise current is then

$$\overline{i_n^2} = \frac{4kT\Delta\nu}{R} \cdot \tag{1.10}$$

The two equivalent circuits are shown in Fig. 1.2. At room temperature the effective noise voltage $\sqrt{\overline{v_n^2}}$ is about 0.13 nV $[\Omega^{-1/2}\,\mathrm{Hz}^{-1/2}]$ and the effective current $\sqrt{\overline{i_n^2}}$ is about 0.13 nA $[\Omega^{1/2}\,\mathrm{Hz}^{-1/2}]$.

The fluctuating thermal noise voltage of a capacitor can be found by considering a closed circuit of a capacitor C in series with a resistor R as shown in Fig. 1.3. The mean square voltage amplitude over the capacity is given by

$$\overline{v_n^2} = \int_0^\infty \frac{4kTR\,\mathrm{d}\nu}{1 + (2\pi\nu CR)^2} = \frac{kT}{C} \cdot \tag{1.11}$$

Since R is not relevant to the result we find that the noise mean square voltage over a capacitor is given by (1.11). It should be noted that the same value for $\overline{v_n^2}$ is found over a resistor connected to a capacitor. Alternatively, one can consider the RC circuit as a low-pass filter having a band width $\Delta\nu = 1/4RC$ for power transmission. Substituting this value of $\Delta\nu$ in (1.9) leads to the same result.

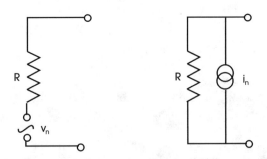

Fig. 1.2. Equivalent circuits

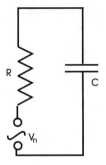

Fig. 1.3. Thermal noise of capacitor

1.3 Shot Noise

Emitted electrons from a thermal cathode or from a photo cathode traveling through a vacuum tube toward the anode produce a current in the external circuit only during their transit time. See Appendix A.1. We assume the total generated current low enough to neglect space charge effects of the electrons in the anode–cathode space so that there are no interactions between the various electrons. Each emitted electron gives a microcurrent pulse. The observed current in the external circuit is then simply the sum of all those randomly generated micropulses. This process also occurs when photons generate electron–hole pairs in a photoconductor or in a photodiode placed between electrodes. The current in the external circuit is only present during the traveling of free electrons to the positive electrode and the holes toward the negative electrode.

The external current due to a random generation of these charge carriers shows as a consequence of the individual pulses uncorrelated fluctuations which are called shot noise. In this section we consider photoemission and assume that each micropulse contains the charge of one electron and has constant duration time. (This is not the case for generation–recombination noise to be treated in Sect. 1.5.) Since the external current can be considered as a flow of electrons that passes a point, one expects by doing a large number of independent observations that the shorter the observation time for counting the number of passing electrons the larger the fluctuations of this number or the larger the shot noise and that by doing observations over large periods the fluctuations and thus the shot noise will approach zero. The analysis is as follows. Let we observe the mean square current fluctuations $\overline{i_n^2}$ of an average current i_0 in the circuit during the time τ_{ob}. The average number of electrons is

$$\overline{n} = \frac{i_0 \tau_{\mathrm{ob}}}{e} . \tag{1.12}$$

The current fluctuation can be expressed as

$$\overline{i_n^2} = \overline{(i - i_0)^2} = \frac{e^2}{\tau_{\mathrm{ob}}^2} \overline{(n - \overline{n})^2} . \tag{1.13}$$

With the assumption that the probability of creating a photoelectron depends on the incident radiation power it is derived in Appendix A.2.4 that for constant radiation power the number n obeys the Poisson statistics with the property

$$\overline{(n - \overline{n})^2} = \overline{n}\,. \tag{1.14}$$

Substituting (1.14) into (1.13) gives

$$\overline{i_n^2} = \frac{i_0 e}{\tau_{ob}}\,. \tag{1.15}$$

1.3.1 Spectral Distribution

In practice it is more useful to express the shot noise in terms of frequency instead of time. For this purpose the Fourier transform relations are used. The (real) function $f(t)$ and its Fourier transform are related by

$$F(\omega) = \int_{-\infty}^{\infty} f(t) e^{-j\omega t}\, dt\,. \tag{1.16}$$

The inverse transform is

$$f(t) = \frac{1}{2\pi} \int_{-\infty}^{\infty} F(\omega) e^{j\omega t}\, d\omega\,. \tag{1.17}$$

Suppose that $f(t)$ is the current in the circuit then the average power over a period T dissipated through a resistor of $1\,\Omega$ is given by

$$P = \frac{1}{T} \int_{-T/2}^{T/2} f^2(t)\, dt = \frac{1}{2\pi T} \int_{-T/2}^{T/2} \left\{ f(t) \int_{-\infty}^{\infty} F(\omega) e^{j\omega t}\, d\omega \right\} dt\,. \tag{1.18}$$

If the current is only present or considered during the time T we find by using (1.16) and (1.17)

$$P = \frac{1}{2\pi T} \int_{-\infty}^{\infty} |F(\omega)|^2\, d\omega = \frac{1}{\pi T} \int_{0}^{\infty} |F(\omega)|^2\, d\omega\,. \tag{1.19}$$

As mentioned earlier the duration of the micropulses of the current flow in the external circuit are related to the time of flow of the generated charge carriers in the detection device. A micropulse current by an electron starting at t_n can be expressed as

$$i_e(t) = e f(t - t_n)\,, \tag{1.20}$$

with the condition

$$\int_{-\infty}^{\infty} f(t - t_n)\, dt = 1\,. \tag{1.21}$$

The total current is then

$$i(t) = e \sum_n f(t - t_n), \qquad (1.22)$$

where t_n is the random starting time of an electron.

Taking the Fourier transform of $i(t)$ we get

$$I(\omega) = e \sum_n e^{-j\omega t_n} \int_{-\infty}^{\infty} e^{-j\omega t} f(t)\, \mathrm{d}t = eF(\omega) \sum_n e^{-j\omega t_n}. \qquad (1.23)$$

The spectral power density $S_i(\omega) = \mathrm{d}P/\mathrm{d}\omega$ of the current $i(t)$ over a resistor of $1\,\Omega$ is according to (1.19)

$$S_i(\omega) = \frac{1}{\pi T} e^2 |F(\omega)|^2 \sum_{n,m} e^{-j\omega(t_n - t_m)} = i_0^2(\omega). \qquad (1.24)$$

Since the times t_n are random and we consider a very large number of electrons, the sum of the terms with $n \neq m$ and $\omega \neq 0$ will for constant production probability of the photoelectrons in average cancel. The summation term for $\omega \neq 0$ becomes equal to the total number of electrons or equal to $\frac{i_0 T}{e}$ where i_0 is the average current of the circuit. We now find for $i_0^2(\omega)$ its average value over T

$$\overline{i_0^2}(\omega) = \frac{1}{\pi} e i_0 |F(\omega)|^2, \qquad (1.25)$$

where the Fourier transform $F(\omega)$ of the micropulse contains the integration over its duration time τ.

The derivation of $\overline{i_0^2}(\omega)$ can be further extended with a description of the micropulse itself or if this is not known by the limitation of the considered bandwidth of the noise spectrum. Let us first consider any micropulse of duration τ for which the considered spectrum is restricted by $\omega\tau \ll 1$. In that case the Fourier transform approaches the unit impulse function i.e.,

$$F(\omega) = \int_0^\tau e^{-j\omega t} f(t)\, \mathrm{d}t \simeq 1. \qquad (1.26)$$

Thus in this case the spectral power density is practically flat, independent on frequency. This shot noise is therefore often considered as white noise. Substituting (1.26) into (1.25) we find the spectral power density of the current fluctuations as

$$\overline{i_0^2}(\omega) = \frac{e i_0}{\pi}. \qquad (1.27)$$

By changing from radial frequency to Hertz frequency (ν) we have to multiply the last expression by 2π and we obtain

$$\overline{i_0^2}(\nu) = 2 e i_0. \qquad (1.28)$$

The current fluctuations or shot noise within a bandwidth B with the condition $2\pi B\tau \ll 1$ becomes

$$\overline{i_n^2} = 2ei_0 B\,. \tag{1.29}$$

Comparing (1.29) and (1.15) it is seen that the relation between the observation time and the bandwidth is given by $\frac{1}{\tau_{ob}} = 2B$. In the following we specify the micropulse for two different situations.

The Charges Move with Constant Speed

Constant speed of created charge carriers by photoionization may occur for instance in a photoconductor or in the high-field region of the junction of a diode. The constant speed during τ gives $f(t) = \frac{1}{\tau}$, where τ is the transit time through the conductor or junction. The Fourier transform of the corresponding micropulse becomes

$$F(\omega) = \int_0^{\tau} e^{-j\omega t}\frac{1}{\tau}\,dt = \frac{\sin(\omega\tau/2)}{\omega\tau/2}e^{-j\omega\tau/2}\,. \tag{1.30}$$

Substituting (1.30) into (1.25) we obtain for the spectral power density of the shot noise

$$\overline{i_0^2}(\omega) = \frac{ei_0}{\pi}\frac{\sin^2(\omega\tau/2)}{(\omega\tau/2)^2}\,. \tag{1.31}$$

This spectrum is shown in Fig. 1.4.

For practical purposes an effective bandwidth $\Delta\nu = \Delta\omega/2\pi$ is calculated for a rectangular spectrum of the same height at the center and of equal area as indicated in Fig. 1.4. The integral $\int_{-\infty}^{\infty}\frac{\sin^2 x}{x^2}\,dx$ is equal to π. This gives an effective half width $\Delta\omega\tau/2 = \pi/2$. The effective maximum noise frequency $\Delta\nu_m$ is then

$$\Delta\nu_m = \frac{1}{2\tau}\,. \tag{1.32}$$

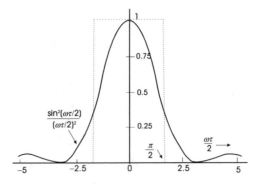

Fig. 1.4. Spectral power density of shot noise. The *dashed line* represents the equivalent rectangular spectrum with half width $\frac{\pi}{2}$

Thus for any noise bandwidth $B < \Delta\nu_m$, the noise is given by (1.29). In practice the value for τ of a hole–electron pair in a diode is in the range of 1–0.01 ns so that $\Delta\nu_m$ is roughly in the range of 1–100 GHz. The detection bandwidth limited by the electronic circuit is mostly much smaller than $\Delta\nu_m$ so that (1.29) remains applicable.

The Charges Move with Constant Acceleration

Constant acceleration of electrons occurs in a vacuum photodiode where a linear potential field is applied between the electrodes so that the velocity of the electrons and thus the current increases linearly with the time of flight of the electrons. The current of the micropulse is then $2et/\tau^2$ where τ is the travel time of the electron from cathode to anode. So we now have $f(t) = 2t/\tau^2$ and the Fourier transform becomes

$$F(\omega) = \frac{2}{(\omega\tau)^2}[(1 + j\omega\tau)\,e^{-j\omega\tau} - 1]. \tag{1.33}$$

Substituting (1.33) into (1.25) results in

$$\overline{i_0^2}(\omega) = \frac{4ei_0}{\pi(\omega\tau)^4}[4\sin^2(\frac{\omega\tau}{2}) + (\omega\tau)^2 - 2\omega\tau\sin\omega\tau]. \tag{1.34}$$

It is found again that $\overline{i_0^2}(0) = ei_0/\pi$.

Plotting the curve of $\overline{i_0^2}(\omega)$ in Fig. 1.5 it is seen to have a broad maximum. The value of $\omega\tau$ for which it reaches its half maximum is $\approx\pi$ so that the maximum noise frequency $\Delta\nu_m$ is again $\Delta\nu_m = 1/2\tau$. Changing again from radial frequency to Hertz frequency we have to multiply (1.34) by 2π. For $\nu < 1/2\tau$ the spectral power density is then again given by (1.28). Consequently the shot noise for the bandwidth B is also given by (1.29).

In conclusion we mention that the spectral power density of the shot noise is determined by the *random* distribution of the micropulses, whereas its maximum frequency is determined by the *duration* of the micropulse which is of course also the maximum frequency response of the detector element.

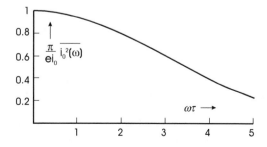

Fig. 1.5. Noise spectrum of rectangular micropulse

1.3.2 Photons

In the earlier analysis we have seen that the current fluctuations are due to the discreteness of the charges. A similar argument applies to a flux of photons. The spontaneously emitted photons of an incoherent radiation source also obey Poisson statistics. The photon fluctuations can then be calculated in a similar way and are obtained simply by replacing the electron charge by the photon energy and the current by the power. We then obtain for the power fluctuations of incoherent radiation having a narrow bandwidth B

$$\overline{\Delta P^2} = 2h\nu PB\,, \tag{1.35}$$

where $h\nu$ is the photon energy ($B \ll \nu$) and P the power of the beam.

For a coherent optical beam with its extremely narrow line width or very long temporal coherence the power fluctuations are negligible. However, the photon current generated by such a beam in any photon detector exhibits nevertheless shot noise as given by (1.29). The derivation of this noise is given in Appendix A.2.4.

1.4 Flicker Noise

Semiconductors and valves show relatively strong noise signals at low frequencies. This noise is usually called flicker noise, $1/f$-noise or excess noise. At low frequencies this noise can be considerably stronger than the shot noise. The observed strong noise signals at low frequencies cannot be fully explained by a description based on the motion of the generated charge carriers. There is more. The search for it has produced many theories based on lattice defects, diffusion of charge carriers, surface contact effects, and impurities. Its origin seems to be very complicated and a full understanding still remains unclear.

Semiempirical studies show a power spectrum more or less inversely proportional to the frequency and quadratic to the current. This frequency dependence remains up till very low frequencies and around 100 Hz it may be comparable with the shot noise. In practice most detection systems operate at frequencies high enough to neglect this type of noise. Therefore, in general it does not limit device performances.

1.5 Generation–Recombination Noise

The previously discussed shot noise is associated with the random generation of identical single charge micropulses. In case of semiconductors the created free carriers increase also the conductivity of the element during the life time of the carriers. As a result the charge of a micropulse initiated by the absorbed photon may be (much) more than that of a single electron. These generated multicharge micropulses are apart from their random distribution not identical because of the life time fluctuations of the carriers. Therefore additional noise is generated [4].

Photoconductors are divided in intrinsic and extrinsic types. In the case of an intrinsic photoconductor the absorbed photon creates a free electron in the conduction band and simultaneously a hole in the valence band. For the extrinsic semiconductor the conduction is produced by the photon absorption at the impurity levels. The photons create either free electrons in the conduction band, the so-called *n-type*, or holes in the valence band of the so-called *p-type*. In general the drift velocity of one type of carrier is much larger than the other one so that in fact the current is given by the dominating type of carrier. Usually the conductivity is mainly by the electrons with their much larger mobility, particularly for intrinsic and n-type extrinsic semiconductors.

Let us consider the drift of the carriers produced by the absorption of photons in a semiconductor crystal connected in series with a battery. See Fig. 1.6. For the optical beam of power P incident on the semiconductor the production rate is $\eta P/h\nu$ electron–hole pairs where η is the quantum efficiency. In steady state the production rate is equal to the recombination rate N/τ_l where N is the number of pairs and τ_l the recombination or life time. Thus we have

$$N = \frac{\eta P \tau_l}{h\nu} . \tag{1.36}$$

Due to the applied field the free carriers drift with constant velocity v between the contacts. Each drifting pair gives rise to an (external) current $i_e = ev/d$ where d is the distance between the contacts with the external leads. The total current using (1.36) becomes

$$i_0 = \frac{e\eta P}{h\nu}\left(\frac{\tau_l}{\tau_d}\right), \tag{1.37}$$

where $\tau_d = d/v$ is the drift time between the contacts. The process can be seen as a carrier, for instance a free electron, that drifts toward the positive contact and leaves the semiconductor. At the same time, because of charge neutrality, a replacement electron enters the semiconductor at the negative contact. This goes on during the life time of the excited charge carrier. The

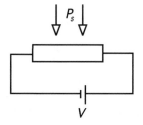

Fig. 1.6. Circuit with semiconductor

micropulse in the external circuit has the duration τ_1 and an effective charge $(\tau_1/\tau_d)e$ per photoinduced charge carrier. The total external current is again the sum of all micropulses that originate from the individually photoinduced charge carriers.

A micropulse starting at t_n can be expressed as

$$i_e(t) = \frac{\tau_1}{\tau_d} e f(t - t_n),\tag{1.38}$$

with the condition given by (1.21). By considering constant drift velocity in the crystal we have a rectangular pulse with $f(t) = \frac{1}{\tau_1}$ for $0 \le t \le \tau_1$ otherwise $f(t) = 0$. The Fourier transform of rectangular pulse is given by (1.30). Substituting this result into (1.25) and now using the effective charge of the pulse equal to $\frac{\tau_1}{\tau_d} e$ we obtain

$$\overline{i_0^2}(\omega) = \frac{e i_0}{\pi} \left(\frac{\tau_1}{\tau_d}\right) \frac{\sin^2(\omega\tau_1/2)}{(\omega\tau_1/2)^2},\tag{1.39}$$

where i_0 is now given by (1.37).

So far the current fluctuations are identical to the shot noise derived in Sect. 1.3 except for the effective charge. The life time of the charge carriers is, however, the result of a spontaneous recombination process and thus it may fluctuate and give rise to additional noise. To include this part of the noise we make the usual assumption that the probability function $F(\tau)$ of the life time is given by

$$F(\tau) = \frac{1}{\tau_1} e^{-\tau/\tau_1},\tag{1.40}$$

so that the average life time is

$$\int_0^\infty \tau F(\tau)\, d\tau = \tau_1.\tag{1.41}$$

Since a life time of a charge carrier is the same as the duration of the corresponding micropulse we can describe the micropulses by the same probability distribution. Then the shot noise produced by the micropulses with duration between τ and $\tau + d\tau$ becomes by using (1.39) and (1.37)

$$d\overline{i_0^2}(\omega) = \frac{e^2\eta P}{\pi h\nu} \left(\frac{\tau}{\tau_d}\right)^2 \frac{\sin^2(\omega\tau/2)}{(\omega\tau/2)^2} \frac{1}{\tau_1} e^{-\tau/\tau_1}\, d\tau.\tag{1.42}$$

Integrating over τ we get the total spectral noise power density

$$\overline{i_0^2}(\omega) = \frac{2eg i_0}{\pi} \left[\frac{1}{1 + (\omega\tau_1)^2}\right],\tag{1.43}$$

where i_0 is given by (1.37) and $g = \tau_1/\tau_d$. Changing to Hertz frequency it becomes

$$\overline{i_0^2}(\nu) = 4eg i_0 \left[\frac{1}{1 + (2\pi\nu\tau_1)^2}\right].\tag{1.44}$$

The term within the brackets indicates the bandwidth limitation due to the carrier life time.

Comparing this g–r noise with the shot noise we see the similarity by noticing that for the g–r noise the "unit of spread" is the charge, ge, of the micropulse whereas for the shot noise it is e. The additional factor 2 in the g–r noise apparently comes from the life time spread of the carriers.

It is seen that the mechanism of producing carriers is not relevant in the derivation of the noise current.[1] The g–r noise is, therefore, also present in the so called "dark" current normally conducted by the thermally excited free carriers and driven by an applied field between the contacts with the external leads. This current consists again of a set of pulses with random arrival times and fluctuating pulse widths because of the statistical behavior of the recombination process. Since the probability of creating thermal free carriers depends on the temperature the dark current consists also of micropulses that obey Poisson statistics provided the temperature is constant. Its noise is therefore also given by (1.44) except that i_0 is now replaced by the dark current i_d induced by the applied field between the contacts and the life time τ_l by the life time of the thermally excited carriers. Calculating the noise the signal and dark currents are in practice often taken as total current in (1.44) assuming a single process for the thermal excitation and the same life time as for the signal carriers. If there are more thermal excitation processes the dark current noise is the sum of the individual contributions. A semiconductor with several g–r processes will make the treatment more complicated.

The dark current noise can also be seen as due to the fluctuations of the resistance, R, of the semiconductor because of the random generation and recombination process of the thermal free carriers. The semiconductor is as a resistor also subjected to the thermal motion of the free carriers colliding with the lattice. Usually the collision time is much shorter than the life time of the carrier so that thermal equilibrium exists with the temperature of the lattice. Thus the semiconductor behaves in addition to the g–r noise also as a resistor with Johnson or thermal noise with an amount given by (1.10).

1.6 Thermal Radiation and Its Fluctuations

According to Planck's radiation law the thermal radiation power at equilibrium temperature T incident on the area A within the small solid angle $d\Omega$ and frequency interval $d\nu$ is given by

$$dP_{\nu,\Omega} = \cos\theta \overline{B} A \, d\nu \, d\Omega \,, \tag{1.45}$$

---

[1] Although the production mechanism is irrelevant the derived Poisson distribution of the micropulses is based on the assumption of constant radiation power. This implies strictly speaking a coherent radiation source. See Appendix A.2.4

where the average brightness $\overline{B}$ is

$$\overline{B} = \frac{2h\nu^3}{c^2(e^{h\nu/kT} - 1)} \qquad (1.46)$$

and θ the angle between the rays and the normal on A. Integrating (1.45) over Ω with $d\Omega = 2\pi \sin\theta\,d\theta$ we obtain for the incident thermal power within the frequency interval $d\nu$ confined within the solid angle Ω_0 of a circular cone with half-angle θ_0

$$dP_\nu = \frac{2\pi h A \sin^2\theta_0}{c^2} \cdot \frac{\nu^3\,d\nu}{e^{h\nu/kT} - 1}. \qquad (1.47)$$

In the ideal case the incident radiation is fully absorbed at the surface (black body). Since in equilibrium the wall temperature remain constant, the surface must emit the same amount. Thus in the ideal case the radiation power emitted in the frequency interval $d\nu$ and area A at temperature T is also given by (1.47). In general the surface is not ideal and it emits less power. Then in order to maintain the constant temperature the incident power must be partly reflected. With a power reflectivity coefficient $\rho(\nu)$ averaged over θ we have for the emission coefficient $\epsilon(\nu)$ the relation

$$\epsilon(\nu) = 1 - \rho(\nu). \qquad (1.48)$$

Integrating (1.47) over the whole spectrum yields at thermal equilibrium for $\theta_0 = \pi/2$ the total incident power on the surface A or emitted average black body power, $\overline{P_t}$, according to the Stefan–Boltzmann law

$$\overline{P_t} = \sigma A T^4, \qquad (1.49)$$

where $\sigma = 2\pi^5 k^4/15c^2 h^3 = 5.67 \times 10^{-8}$ [W m^{-2} K^{-4}].

Let us now consider the average thermal power $\overline{P}$ incident normal to a detector surface A within the small solid angle $\Delta\Omega$ and bandwidth $\Delta\nu$. Taking $\cos\theta \approx 1$ and assuming $\Delta\nu$ is small compared to kT/h so that the considered thermal energy per unit frequency is constant we obtain from (1.45)

$$\overline{P} = \frac{\Delta\Omega A}{\lambda^2} \frac{2h\nu}{e^{h\nu/kT} - 1}\Delta\nu. \qquad (1.50)$$

The quantized average energy of a single frequency mode is equal to

$$\overline{E}_{h\nu} = \frac{h\nu}{e^{h\nu/kT} - 1}. \qquad (1.51)$$

The factor 2 in (1.50) refers to the two independent polarizations of the field. We should keep in mind that $\Delta\nu$ is the selected optical bandwidth which is always much larger than the electronic bandwidth B of the detection system.

The incident thermal radiation can be derived by any set of orthogonal field functions that completely fills the space bounded by the plane containing the area A. If the radiation originates from a distance large enough to receive all wavefronts of the radiation at A parallel, the field components on A are coherent. Then, considering (1.50), the minimum space of photons, i.e., a single spatial mode, is bounded by the condition

$$\Delta\Omega A \approx \lambda^2 \,. \tag{1.52}$$

In case the incident radiation falls within an area–angle product larger than λ^2 its field is build up of several spatial modes. The number is given by

$$N_m \approx \frac{\Delta\Omega A}{\lambda^2} \,. \tag{1.53}$$

Each spatial mode contains many frequency modes. The power within a single spatial mode for one polarization is derived in Appendix A.4 and is given by

$$\overline{P}_m = \overline{E}_{h\nu}\Delta\nu \,. \tag{1.54}$$

By substituting (1.53) and (1.51) into (1.50) we get

$$\overline{P} = 2N_m\overline{E}_{h\nu}\Delta\nu \,. \tag{1.55}$$

The noise associated with this thermal radiation consists of two parts. One part is due to the quantization of the radiation. The radiation may be regarded as a stream of fluctuating photons. The thermally emitted photons at constant temperature with their finite lifetimes obey Poisson statistics. The noise power for a detection bandwidth B corresponds to shot noise, similar to what has been described for the random flow of charge particles in Sect. 1.3. Looking at (1.29) we have to replace for the similarity the current i_0 by the radiation power $\overline{P}$ and the charge e by the photon energy $h\nu$. Applying (1.35) we find for this part of the power fluctuations

$$\overline{\Delta P_{\mathrm{sh}}^2} = 2h\nu\overline{P}B \,. \tag{1.56}$$

In practice the band width B of the detection system is always much smaller than the optical bandwidth $\Delta\nu$ of the selected thermal radiation.

The other part results from the fluctuations of amplitude and phase. Within a single spatial mode the instantaneous radiation field results from the concerted action of a very large number of independent emitters. Therefore, from a statistical point of view the central limit theorem is appropriate and the resulting radiation field amplitude and its phase within a spatial mode are Gaussian processes. Their mutually independent fluctuations can be evaluated by considering the field as composed of two components in an arbitrary rectangular coordinate system and then apply the Gaussian process to each component. The Gaussian distribution of the field component v_x in the x-direction with the probability $F(v_x)\,\mathrm{d}v_x$ for having its value between v_x and $v_x + \mathrm{d}v_x$ at any time is given by

$$F(v_x)\,\mathrm{d}v_x = \frac{1}{\sqrt{2\pi}\sigma}\,\mathrm{e}^{-v_x^2/2\sigma^2}\,\mathrm{d}v_x \tag{1.57}$$

Similarly for $F(v_y)$ we have

$$F(v_y)\,\mathrm{d}v_y = \frac{1}{\sqrt{2\pi}\sigma}\,\mathrm{e}^{-v_y^2/2\sigma^2}\,\mathrm{d}v_y\,. \tag{1.58}$$

The probability $F(v_x + v_y)\,\mathrm{d}v_x\,\mathrm{d}v_y$ of finding the x-component between v_x and $v_x + \mathrm{d}v_x$ and the y-component between v_y and $v_y + \mathrm{d}v_y$ is then

$$F(v_x + v_y)\,\mathrm{d}v_x\,\mathrm{d}v_y = \frac{1}{2\pi\sigma^2}\,\mathrm{e}^{-(v_x^2+v_y^2)/2\sigma^2}\,\mathrm{d}v_x\,\mathrm{d}v_y\,. \tag{1.59}$$

Changing to the circular components v and φ with $v^2 = v_x^2 + v_y^2$ and $\mathrm{d}v_x\,\mathrm{d}v_y = v\mathrm{d}\,v\mathrm{d}\varphi$ and integrating over φ we find the field probability of v

$$F(v)\,\mathrm{d}v = \frac{1}{\sigma^2}\,\mathrm{e}^{-v^2/2\sigma^2}\,v\,\mathrm{d}v\,. \tag{1.60}$$

The power P_m within a spatial mode is proportional to v^2. The average power $\overline{P_m}$ is then obtained by multiplying the last equation by v^2 and substituting $P_m = \alpha v^2$. We obtain

$$\overline{P_m} = \int_0^\infty \frac{P_m}{2\alpha\sigma^2}\,\mathrm{e}^{-P_m/2\alpha\sigma^2}\,\mathrm{d}P_m = 2\alpha\sigma^2\,. \tag{1.61}$$

Thus the radiation power probability distribution $F(P_m)$ of a single spatial mode, which may contain a set of frequency modes, is given by

$$F(P_m) = \frac{1}{\overline{P_m}}\,\mathrm{e}^{-P_m/\overline{P_m}} \tag{1.62}$$

which is called the Rayleigh distribution.

The power spread of a single spatial mode is given by

$$\overline{\Delta P_m^2} = \overline{\left(P_m - \overline{P_m}\right)^2} = \overline{P_m^2} - \overline{P_m}^2\,. \tag{1.63}$$

Using (1.62) we find $\overline{P_m^2}$ equal to $2\overline{P_m}^2$ so that

$$\overline{\Delta P_m^2} = \overline{P_m}^2\,. \tag{1.64}$$

Taking the sum of the energy spreads of all spatial modes we just multiply the last equation by $2N_m$ because the spatial mode are independent from each other. We obtain

$$\overline{\Delta P_{ray}^2} = 2N_m\overline{P_m}^2\,. \tag{1.65}$$

With the aid of (1.54) and (1.55) we get

$$\overline{\Delta P_{ray}^2} = \overline{PE}_{h\nu}\Delta\nu \,. \tag{1.66}$$

The last expression contains the noise power spread over its full optical spectrum $\Delta\nu$. We are now interested to know its frequency distribution because the detector with its much smaller bandwidth B receives only a small part of it. We assume $\Delta\nu$ to be small compared to kT/h so that the considered thermal power per unit frequency is constant. Following a classical description we note that the power fluctuations within a spatial mode correspond to beat frequencies between the field components of the frequency modes. Since the bandwidth of the selected radiation is $\Delta\nu$ the radiation power has components with beat frequencies ν_b ranging from zero to $\Delta\nu$. The number of beat components is highest for $\nu_b = 0$ and the power of these beat components decreases linearly with $(1 - (\nu_b/\Delta\nu))$ to reach zero for $\nu_b = \Delta\nu$. This is indicated in Fig. 1.7. It is seen from this figure that the noise content for a spectrum with $B \ll \Delta\nu$ is the fraction $2B/\Delta\nu$ of the total noise. Using (1.66) the Rayleigh noise within the bandwidth B becomes

$$\overline{\Delta P_{ray}^2} = 2B\overline{PE}_{h\nu} \,. \tag{1.67}$$

The total noise within the bandwidth B is the sum of parts given by, respectively, (1.56) and (1.67) or

$$\overline{\Delta P^2} = 2h\nu\overline{P}\left(1 + \frac{1}{e^{h\nu/kT} - 1}\right)B \,, \tag{1.68}$$

where we used (1.51).

It is interesting that the result given by (1.68) can also be derived straightforward from statistical thermodynamics. This is done by starting from the partition function, $Z_{h\nu}$, of a radiation mode with photon energy $h\nu$ given by

$$Z_{h\nu} = \sum_{n=0}^{\infty} e^{-nh\nu/kT} = \frac{1}{1 - e^{-h\nu/kT}} \,. \tag{1.69}$$

The average energy, $\overline{E}_{h\nu} = \frac{1}{Z_{h\nu}} \sum_{n=0}^{\infty} nh\nu\, e^{-nh\nu/kT}$, is given by

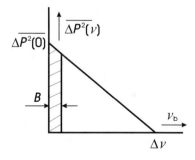

Fig. 1.7. Spectral distribution of thermal noise

$$\overline{E}_{h\nu} = \left[1 - e^{-h\nu/kT}\right] kT^2 \frac{d}{dT} Z_{h\nu} = \frac{h\nu}{e^{h\nu/kT} - 1} \, . \tag{1.70}$$

The average square of the energy, $\overline{E_{h\nu}^2} = \frac{1}{Z_{h\nu}} \sum_{n=0}^{\infty} (nh\nu)^2 \, e^{-nh\nu/kT}$, is calculated similarly by

$$\overline{E_{h\nu}^2} = \left[1 - e^{-h\nu/kT}\right] kT^2 \frac{d}{dT} \left[\frac{\overline{E}_{h\nu}}{1 - e^{-h\nu/kT}}\right] = (h\nu)^2 \frac{e^{h\nu/kT} + 1}{\left(e^{h\nu/kT} - 1\right)^2} \, . \tag{1.71}$$

Further we have $\overline{\Delta E_{h\nu}^2} = \overline{\left(E_{h\nu} - \overline{E}_{h\nu}\right)^2} = \overline{E_{h\nu}^2} - \overline{E}_{h\nu}^2$. Substituting (1.70) and (1.71) we get

$$\overline{\Delta E_{h\nu}^2} = h\nu \overline{E}_{h\nu} \left(1 + \frac{1}{e^{h\nu/kT} - 1}\right) \, . \tag{1.72}$$

The relation between the energy of a single frequency mode and the power of a spatial mode is derived in Appendix A.4. The fluctuations of the total power of the considered beam is the sum of the fluctuations of each spatial mode or by using (1.55)

$$\overline{(\Delta P^2)}_{total} = \overline{(P - \overline{P})^2} = 2N_m (\Delta\nu)^2 \overline{\left(E_{h\nu} - \overline{E}_{h\nu}\right)^2} = 2 (\Delta\nu)^2 N_m \overline{\Delta E_{h\nu}^2} \, , \tag{1.73}$$

where the number of independent spatial modes N_m is multiplied by 2 because of the two independent polarizations. Substituting (1.72) and (1.55) into (1.73) we get

$$\overline{(\Delta P^2)}_{total} = h\nu \overline{P} \left(1 + \frac{1}{e^{h\nu/kT} - 1}\right) \Delta\nu \, , \tag{1.74}$$

which contains the noise power spread over $\Delta\nu$. Since we are interested in the contributions within the band $B \ll \Delta\nu$ we follow the previous discussion to derive (1.67) and replace $\Delta\nu$ by $2B$. We obtain in agreement with (1.68)

$$\overline{\Delta P^2} = 2h\nu \overline{P} \left(1 + \frac{1}{e^{h\nu/kT} - 1}\right) B \, . \tag{1.75}$$

In the optical region where $h\nu/k$ is much larger than T the second term within the brackets is negligible, but this is not the case for thermal radiation.

1.7 Temperature Fluctuations of Small Bodies

The energy of a body is always in interaction with its surroundings and its transfer occurs by statistical processes of radiation, convection, and conduction. Even at equilibrium its value will fluctuate randomly about a mean value. The question is how large are those fluctuations and how do they depend on physical quantities. An elegant way to answer these questions is to apply statistical thermodynamics and derive the formula for the probability that a system has an energy E.

　　Let us have a large number of identical systems that can assume energy values $E_1, E_2, E_3, \dots$ and which are all in thermal heat exchange with a large temperature bath kept at constant temperature T. According to Boltzmann the probability that a system has an energy E_i is $Ae^{-E_i/kT}$. The sum of all probabilities must be one, so that $\sum_i Ae^{-E_i/kT} = 1$. The average energy, $\overline{E}$, is obtained by

$$\overline{E} = \sum_i AE_ie^{-E_i/kT} = \frac{\sum_i E_ie^{-E_i/kT}}{\sum_i e^{-E_i/kT}}. \tag{1.76}$$

Taking the temperature derivative of $\overline{E}$

$$\frac{d\overline{E}}{dT} = \frac{1}{kT^2}\left[\frac{\sum_i E_i^2e^{-E_i/kT}}{\sum_i e^{-E_i/kT}} - \left(\frac{\sum_i E_ie^{-E_i/kT}}{\sum_i e^{-E_i/kT}}\right)^2\right] = \frac{1}{kT^2}\left[\overline{E^2} - \overline{E}^2\right], \tag{1.77}$$

which is the heat capacity of the system.

We now apply this result to a small body and describe thereby the heat content by its temperature so that energy fluctuations will be interpreted as temperature fluctuations according to $E - \overline{E} = C_{th}(T - \overline{T})$ where C_{th} is the heat capacity. Then

$$\overline{E^2} - \overline{E}^2 = \overline{(E - \overline{E})^2} = C_{th}^2\overline{(T - \overline{T})^2} = C_{th}^2\overline{\Delta T^2}. \tag{1.78}$$

Substituting (1.78) into (1.77) we get

$$\overline{\Delta T^2} = \frac{kT^2}{C_{th}}. \tag{1.79}$$

Next we want to describe the spectral density of the temperature fluctuations. For that purpose we realize that frequency fluctuations are damped by the thermal slowness of the system which has a time constant $\tau_{th} = C_{th}/\lambda$ where λ is the thermal conductance. It behaves analogously to a RC circuit in electronics. This is obvious if we relate ΔT to voltage, C_{th} to electrical capacity and λ to electrical conductivity. Since we describe mean-square fluctuations we obtain from the analogy the following frequency (f) dependence

$$\overline{\Delta T^2}(f) = \frac{\overline{\Delta T^2}(0)}{1 + (2\pi f\tau_{th})^2} \tag{1.80}$$

Integrating the last equation over all frequencies we get again the total value given by (1.79). From this we find $\overline{\Delta T^2}(0) = 4kT^2/\lambda$ and write (1.80) as

$$\overline{\Delta T^2}(f) = \frac{4kT^2}{\lambda}\frac{1}{1 + (2\pi f\tau_{th})^2}. \tag{1.81}$$

Integrating (1.81) over a bandwidth B much smaller than the reciprocal thermal time we get

$$\overline{\Delta T^2} = \frac{4kT^2B}{\lambda}\frac{1}{1 + (2\pi f\tau_{th})^2}. \tag{1.82}$$

These temperature fluctuations limit the minimum detectable power of thermal detectors.

1.7.1 Absorption and Emission Fluctuations

Let us consider the situation that a black body is in equilibrium with the thermal radiation of the surrounding and that all energy transfer is by radiation only. The thermal fluctuations of the body are then related to both emission and absorption fluctuations. A small change of radiation transfer ΔP, either by absorption or emission, results in a small temperature change ΔT of the body. The relation between ΔP and ΔT derived from (1.49) yields

$$\lambda = \frac{\mathrm{d}P}{\mathrm{d}T} = 4\sigma AT^3 \,. \tag{1.83}$$

The mean square power fluctuations near $f = 0$ within the bandwidth B are obtained by substituting (1.83) into (1.82). We find

$$\overline{\Delta P^2} = 16AB\sigma kT^5 \,. \tag{1.84}$$

The fluctuations of the incident absorbing radiation in the case of no reflection are obtained by integrating (1.68) over the full spectrum. This is done by substituting for $\overline{P}$ the expression $\mathrm{d}P_\nu$ from (1.47) and integrating over ν. We obtain

$$\overline{\Delta P_{\mathrm{abs}}^2} = \frac{4\pi AB h^2 \sin^2 \theta_0}{c^2} \int_0^\infty \frac{\nu^4 e^{h\nu/kT}}{\left(e^{h\nu/kT} - 1\right)^2} \,\mathrm{d}\nu \tag{1.85}$$

or

$$\overline{\Delta P_{\mathrm{abs}}^2} = \frac{16\pi^5 AB k^5 T^5 \sin^2 \theta_0}{15 c^2 h^3} = 8\sin^2 \theta_0 AB\sigma kT^5 \,, \tag{1.86}$$

where we have substituted $\sigma = 2\pi^5 k^4/15c^2 h^3$. It is seen that for the total incident radiation with $\theta_0 = \pi/2$ we find just one-half of what is obtained for the total fluctuations given by (1.84). This can be understood by the fact that we have considered so far only the incident radiation. If the area is in thermal equilibrium with the radiation field, there will be an equal amount of power fluctuations emitted by the area A so that the total fluctuations are the same as derived by (1.84).

2

Signal–Noise Relations

The noise signals as discussed in Chap. 1 are always inherently connected to the detection process or they are a consequence of the physical nature of the input signal itself. In fact the noise pollutes the signal and a low-power signal may be even obscured by the noise. This means that for any detection system, there is a minimum signal power below which detection fails and for signals well above the noise power, there is always an inaccuracy set by the present noise. Thus the noise limits the accuracy and sensitivity of a detection system.

In general, the noise power that is mixed with the observed signal is quantitatively described by the so-called noise-equivalent-power (NEP), which is the input power that gives a signal equal to that of the noise. To quantify the accuracy and sensitivity of a detection system the signal-to-noise ratio (S/N) is used. This ratio refers usually to the detector output powers produced by signal and noise, respectively, although it may sometimes refer to the signal and noise output voltages. In practice the NEP-value is often calculated for each individual noise source and the total NEP is obtained by the quadratic combination of these NEPs. In this way the relative contributions and the dominating noise source(s) are indicated.

The minimum detectable power of a detector is often indicated by its detectivity D rather than its NEP. The mutual relation is simply

$$D = \frac{1}{\text{NEP}} \, .$$
(2.1)

Next one quotes quite often the specific detectivity D^* which is useful for the comparison of the performances of different detectors. Because the NEP is in general proportional to the square root of the product of the detector area and the frequency bandwidth, the specific detectivity is defined as

$$D^* = \frac{\sqrt{AB}}{\text{NEP}}$$
(2.2)

2.1 Signal Limitation

Radiation detectors like photodetectors and thermal detectors convert the absorbed radiation into electrical output signals. They are often called square law detectors because they respond to the intensity of radiation which is proportional to the square of its field. (The time constants of these detectors are too large to respond to optical or infrared frequencies.) The general expression for the conversion of the power P_s with optical frequency ν into a photon current i_s is

$$i_s = \frac{e\eta P_s}{h\nu} , \tag{2.3}$$

where $h\nu$ is the photon energy, $P_s/h\nu$ the number of photons per unit time interval, e the electron charge, and η the conversion or quantum efficiency i.e., the fraction of incident photons that are converted into electrons. If we are dealing with constant incident beam power the statistical distribution of the generated electrons is a Poisson distribution as derived in Appendix A.2.4. The noise current is given by

$$\overline{i_n^2} = 2ei_s B , \tag{2.4}$$

where B is the bandwidth of the detection system. To reduce the noise the bandwidth B is made as narrow as possible to just transmit the signal. The signal-to-noise ratio becomes with (2.3) and (2.4)

$$\frac{S}{N} = \frac{i_s^2}{i_n^2} = \frac{\eta P_s}{2h\nu B} . \tag{2.5}$$

Since the signal noise itself is considered the corresponding NEP is called signal limited and is derived from $S/N = 1$. We obtain

$$\mathrm{NEP_{SL}} = \frac{2h\nu B}{\eta} . \tag{2.6}$$

This is the minimum detectable signal power for the detection bandwidth B. This minimum can also be understood by analyzing it differently and looking at the time space instead of the frequency space. For that purpose we replace $2B$ by $1/\tau$ according to (1.32), where τ is the observation time during which the photons are counted. Thus we now obtain $\mathrm{NEP_{SL}} = h\nu/\eta\tau$ or on the average one photon in the observation time which is of course the minimum detectable power or detectable change of power. In the case we are dealing with g–r noise it is readily found that $\mathrm{NEP_{SL}}$ is a factor two larger.

2.2 Background Limitation

Any detection system deals with background radiation and thermal fluctuations that mix with the observed signal. Because background fluctuations depend strongly on the ambient temperature a considerable reduction of this

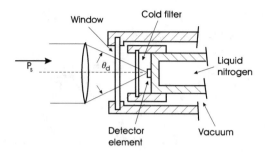

Fig. 2.1. Liquid nitrogen cooled detector with reduced incident thermal radiation

background radiation is obtained by cooling the detector and its housing. A schematic construction of a liquid nitrogen cooled detector element is shown in Fig. 2.1. However, it is unavoidable that nevertheless a part of the thermal radiation from the outside environment enters by the signal collecting aperture of the device. The field of view is generally restricted by the entrance aperture containing the signal focusing optics. By also cooling the entrance window, only background radiation from the outside that falls within the focusing cone with angle θ_d, as depicted in the figure, will reach the detector element. For detecting low power, narrow-band, infrared radiation the effect of background radiation can be further reduced by incorporating a cooled narrow-band filter (cold filter) in front of the detector element as shown in the figure. It restricts the transmitted ambient radiation to a narrow band.

In the following we consider the incident background radiation in the absence of a cold filter. The brightness of radiation is defined as the radiation power per unit radiation frequency and per unit solid angle passing perpendicular a unit area. It is a well-known theorem of imaging that if radiation is transported by an optical system its brightness cannot be greater than the original. If the losses due to absorption and partial reflection at lens surfaces or to incomplete reflection at mirror surfaces are neglected the brightness of the image is equal to that of the object. We apply this theorem to the thermal radiation imaged on the detector and neglect losses. By calculating the detected background radiation entering from the outside through the aperture we take the brightness given by (1.46) where T is the outside temperature. The radiation extends over the cone angle θ_d and reaches the detector surface A. The detected radiation is then obtained by integrating (1.45)

$$P_{\mathrm{B}} = \int_0^{\theta_d} \int_0^\infty 2\pi \overline{B} A \cos\theta \sin\theta \, d\nu \, d\theta = 2\pi \sin^2\left(\frac{\theta_d}{2}\right) \frac{A}{c^2} \int_0^\infty \frac{h\nu^3}{e^{h\nu/kT} - 1} \, d\nu \tag{2.7}$$

or

$$P_{\mathrm{B}} = \sin^2\left(\frac{\theta_d}{2}\right) \sigma T^4 A \,, \tag{2.8}$$

where $\sigma = 2\pi^5 k^4 / 15 c^2 h^3 = 5.67 \times 10^{-8}\,\mathrm{Wm^{-2}K^{-4}}$ and θ_d the cone angle.

In case the detector element is not cooled and in thermal equilibrium with its surroundings it may be treated as a black body (no reflection) with area A. Its mean square thermal fluctuations are then given by (1.84). If the noise of the detector is only due to these thermal fluctuations, the system is said to be background limited. NEP by definition is equal to

$$\text{NEP}_{\text{BL}} = 4\sqrt{AB\sigma kT^5}.\tag{2.9}$$

The corresponding specific detectivity given by (2.2) is then called ideal detectivity because it gives for incoherent detection[1] its ultimate value. Thus if the background is in thermal equilibrium with the detector element we get by substituting (2.9) into (2.2) for the ideal detectivity

$$D_i^* = \frac{1}{4\sqrt{\sigma kT^5}},\tag{2.10}$$

which for $T=300$ K gives $D_i^* = 1.8 \times 10^{10}\,\text{W}^{-1}\text{cm}\,\text{Hz}^{1/2}$. This value is the upper limit of a thermal detector in equilibrium with the background.

In case the detector housing is cooled at cryogenic temperature having negligible brightness $\overline{B}$ and the signal is collected by a solid angle $\Delta\Omega$ with cone angle θ_d as indicated in Fig. 2.1, the fluctuations in the absence of reflections are given by (1.86) where $\theta_0 = \theta_d/2$. Neglecting again all other noise sources the corresponding NEP becomes

$$\text{NEP}_{\text{BL}} = \sin\left(\frac{\theta_d}{2}\right)\sqrt{8AB\sigma kT^5}.\tag{2.11}$$

2.2.1 Ideal Detection

Unlike thermal detectors, the performance of photon detectors is not independent of wavelength. A photon detector responds to photons with energy $h\nu$ larger than the minimum energy $h\nu_0$ required to excite the electronic transition.

We consider the ideal detector for which it is assumed that the detector is cooled at cryogenic temperature with negligible brightness, each incident photon with energy greater than $h\nu_0$ produces one photon electron and the incident background radiation is again limited by the aperture with cone angle θ_d as indicated in Fig. 2.1. The mean square fluctuations in the background radiation to which the photon detector responds are described by (1.85) where the integration is now between the limits ν_0 and ∞.

The incident background radiation can also be written in terms of photon rate n times the corresponding photon energy $h\nu$. The fluctuations of n are obtained by substituting for $\overline{P}$ in (1.68) the expression dP_ν from (1.47),

---

[1] Incoherent or direct detection in contrast to coherent or heterodyne detection.

dividing both sides of (1.68) by $(h\nu)^2$ and then integrating the result over the frequency spectrum. We find by substituting $x = h\nu/kT$

$$\overline{\Delta n^2} = \frac{4\pi AB\,(kT)^3 \sin^2(\theta_d/2)}{h^3 c^2} \int_{x_0}^{\infty} \frac{x^2 e^x}{(e^x - 1)^2}\,\mathrm{d}x\,, \qquad (2.12)$$

where $x_0 = h\nu_0/kT$. (This result is also obtained by dividing the integrand of (1.85) by $(h\nu)^2$.)

Although strictly speaking the number n is obtained by dividing both sides of (1.47) by $h\nu$ and then by integrating the result over the frequency spectrum we use instead the relation $\overline{\Delta n^2} = 2Bn$ where $\overline{\Delta n^2}$ is given by (2.12). This gives, especially for values of $\nu_0 < kT/h$, a slightly different result. When we apply this value of n to calculate the background current i_b by $i_b = en$ we derive with this value of i_b the correct current fluctuations in the same way as will be done for the signal and dark currents. The background current fluctuations become

$$\overline{\Delta i_b^2} = 2ei_b B = 2e^2 nB = e^2 \overline{\Delta n^2}\,. \qquad (2.13)$$

For calculating the NEP we have the condition that the number of signal quanta must be equal to the root mean square fluctuations of the incident photon rate from the background. The corresponding power will be smallest for the smallest energy $h\nu_0 = h\nu_s = hc/\lambda_s$ where λ_s is the signal wavelength. For the ideal detector we consider $\eta = 1$ and obtain for the background limited NEP

$$\mathrm{NEP}_{\mathrm{BL}} = \frac{hc}{\lambda_s} \left[\frac{4\pi AB\,(kT)^3 \sin^2(\theta_d/2)}{h^3 c^2} \int_{x_s}^{\infty} \frac{x^2 e^x}{(e^x - 1)^2}\,\mathrm{d}x \right]^{1/2}. \qquad (2.14)$$

Using (2.2) the ideal detectivity for photon detectors becomes

$$D_i^* = \left[\frac{4\pi\,(kT)^3 \sin^2(\theta_d/2)}{h\lambda_s^2} \int_{x_s}^{\infty} \frac{x^2 e^x}{(e^x - 1)^2}\,\mathrm{d}x \right]^{-1/2} \qquad (2.15)$$

or

$$D_i^* = 1.3 \times 10^{11} \left(\frac{300}{T} \right)^{5/2} \frac{1}{\sin(\theta_d/2)} \frac{1}{x_s} \left[\int_{x_s}^{\infty} \frac{x^2 e^x}{(e^x - 1)^2}\,\mathrm{d}x \right]^{-1/2}, \qquad (2.16)$$

with D_i^* expressed in $\mathrm{W}^{-1}\,\mathrm{cm}\,\mathrm{Hz}^{1/2}$.

In Fig. 2.2 the ideal detectivity D_i^* is plotted as a function of the wavelength (in µm) for $T = 300\,\mathrm{K}$ and $\theta_d = 180°$ which corresponds to 2π steradian field of view. Dealing with different temperatures the curves are still applicable by pointing in the diagram at a wavelength that is $T/300$ times the considered wavelength and then multiplying the corresponding value of

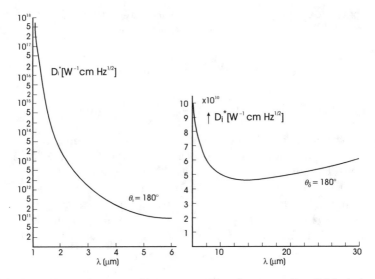

Fig. 2.2. Ideal detectivity for $T = 300\,\mathrm{K}$ and 2π steradian field of view

D_i^* by $(300/T)^{5/2}$. In case the photon detector generates g–r noise the found value of D_i^* must by dividing by $\sqrt{2}$ (see Sect. 1.5).

Next we take the quantum efficiency η of the detector into account. The effective rate of photons is now ηn. This is also the case for the effective fluctuations on the detector which are now $\eta \overline{\Delta n^2}$. So the $\mathrm{NEP_{BL}}$ becomes

$$\mathrm{NEP_{BL}} = \frac{hc}{\lambda_\mathrm{s}} \left(\frac{\overline{\Delta n^2}}{\eta} \right)^{1/2}, \qquad (2.17)$$

where $\overline{\Delta n^2}$ is given by (2.12). The shot noise current of a photon detector generated by the background radiation is then given by

$$\overline{i_\mathrm{n}^2} = \left(\frac{e\eta}{h\nu_\mathrm{s}} \right)^2 \mathrm{NEP_{BL}^2} = 2\eta e^2 Bn = 2ei_\mathrm{b}B, \qquad (2.18)$$

where we have used (2.13).

In case the photon detector generates g–r noise the corresponding $\mathrm{NEP_{BL}}$ is a factor $\sqrt{2}$ larger and consequently the corresponding shot noise becomes

$$\overline{i_\mathrm{n}^2} = 2 \left(\frac{ge\eta}{h\nu_\mathrm{s}} \right)^2 \mathrm{NEP_{BL}^2} = 4\eta \left(ge \right)^2 Bn = 4gei_\mathrm{b}B, \qquad (2.19)$$

where $i_\mathrm{b} = \eta gen$. We note that the factor 2 in (2.19) comes from the statistical spread of the recombination time as was pointed out in Sect. 1.5.

2.3 Johnson Noise

If only Johnson noise is considered the corresponding $\mathrm{NEP}_{\mathrm{AL}}$ is usually called amplifier limited. For the S/N-value we apply (1.10) and (2.3) and obtain

$$\frac{S}{N} = \left(\frac{e\eta P_{\mathrm{s}}}{h\nu}\right)^2 \frac{R}{4kTB} \tag{2.20}$$

so that

$$\mathrm{NEP}_{\mathrm{AL}} = \frac{h\nu}{e\eta} \sqrt{\frac{4kTB}{R}} . \tag{2.21}$$

2.4 Dark Current Noise

It is mostly unavoidable that detectors have bias currents or generate currents due to very different processes than the photo excitation process. These currents interfere with the signal current. Typical examples are thermionic emission and cosmic rays in diodes or the currents due to the ohmic resistance of semiconductor photodetectors. In general, these currents are composed of a strong dc component with fluctuations superimposed on it. These disturbing currents are also present in the absence of signal radiation and are therefore called "dark" currents. Similar to the background radiation signal the dc component of the dark currents can be eliminated by operating the detector with its signal in the alternating mode and using a blocking capacitor in the circuit. If these dark currents obey the Poisson statistics the fluctuations which (together with the signal current) pass the blocking capacitor give a noise current described by (1.29) or

$$\overline{i_{\mathrm{n}}^2} = 2ei_{\mathrm{d}}B , \tag{2.22}$$

where i_{d} is the dark current and B the electronic bandwidth of the detector device. Using (2.3) the S/N-value becomes

$$\frac{S}{N} = \frac{\left(\frac{e\eta P_{\mathrm{s}}}{h\nu}\right)^2}{2ei_{\mathrm{d}}B} \tag{2.23}$$

and the dark current limited $\mathrm{NEP}_{\mathrm{DL}}$ is then

$$\mathrm{NEP}_{\mathrm{DL}} = \frac{h\nu}{e\eta} \sqrt{2ei_{\mathrm{d}}B} . \tag{2.24}$$

In case the dark current generates g–r noise i_d must be replaced by $2i_{\mathrm{d}}/g$.

2.5 Noise and Sensitivity

Similar treatments as above can be given for any additional random noise source and for each noise source the NEP can be derived. Since the average quadratic fluctuations of all these independent sources can be simply added to obtain the total we get

$$\mathrm{NEP} = \left[\sum_i \mathrm{NEP}_i^2 \right]^{1/2} \tag{2.25}$$

and consequently the final S/N-value becomes

$$\frac{S}{N} = \frac{P_s^2}{(\mathrm{NEP})^2} \, . \tag{2.26}$$

We note that the NEP is proportional to the square root of the bandwidth B. This follows from the fact that the output noise power is proportional to the bandwidth and the output signal power proportional to the square of the input radiation power. These devices are therefore often called square-law detectors.

The sensitivity of a detector is also directly related to its NEP and S/N-ratio. With sensitivity is meant the minimum increase of input power that can be detected. The higher the sensitivity the smaller the observed change of input power. In other words, it is the change of power ΔP that rises the signal above its noise ripple or mathematically

$$(P_s + \Delta P_s)^2 > P_s^2 + \mathrm{NEP}^2 \tag{2.27}$$

or for $P_S \gg \mathrm{NEP}$

$$\Delta P_s > \frac{\mathrm{NEP}^2}{2 P_s} \tag{2.28}$$

Thus the sensitivity is proportional to the square of the NEP and the relative sensitivity $\Delta P_s / P_s = N/2S$ is inversely proportional to the signal-to-noise ratio.

2.6 Amplifier Noise and Mismatching

For weak detector signals it is often necessary to raise the power level by passing it through a linear amplifier. In practice it is not feasible to have an amplifier that simply gives a linear enlarged copy of both input signal and its noise. The operational characteristics of the amplifier may to some extend distort the input waveform and add excess noise to the output and thereby deteriorating the signal-to-noise ratio. A desirable amplifier produces a good

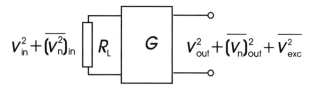

$$V_{\text{in}}^2 + \overline{\left(V_n^2\right)}_{\text{in}} \quad \boxed{R_L} \quad \boxed{G} \quad V_{\text{out}}^2 + \overline{\left(V_n\right)}_{\text{out}}^2 + \overline{V_{\text{exc}}^2}$$

Fig. 2.3. Amplifier noise taken into account by the effective temperature

replica of the input signal with a minimum of excess noise to it. Usually an operational amplifier is installed which has high frequency response with good linearity. This is discussed in Sect. 7.1.

The additional noise of the amplifier is indicated by its so-called *noise figure* . It is in fact a parameter that describes the ratio of the amplifier noise to the thermal noise of the input source resistance. Its use is very practical and facilitates the computation of the output signal-to-noise ratio with the inclusion of the amplifier noise.

The input of an amplifier is coupled to a load resistance R_L as shown in Fig. 2.3. The input impedance of the amplifier is much larger than R_L. At the input the source signal is only mixed with the Johnson noise from the load resistance which is at the conventional temperature $T_0 = 290\,\text{K}$.

The gain G of the amplifier is the ratio of the output voltage to the input voltage i.e., $G = V_{\text{out}}/V_{\text{in}}$. It is often expressed in terms of decibels equal to $10 \log G$. Thus $10\,\text{dB}$ means a voltage gain of 10 and a power gain of 100. The noise at the output of the amplifier is the amplified Johnson noise and the excess noise of the amplifier itself. For this situation we define the amplifier noise figure F as the ratio of the S/N-value at the input to that at the output or

$$F = \frac{(S/N)_{\text{in}}}{(S/N)_{\text{out}}} . \tag{2.29}$$

Usually F is also quoted in decibels by $10 \log F$. A perfect amplifier has a noise figure of 0 dB while, for instance, one that degrades the S/N-value of the signal source by a factor of two has a 3 dB noise figure. In general, the parameters G and F are functions of frequency. An example of a diagram is shown in Fig. 2.4, where F depends on both frequency and load impedance.

As the input impedance is much larger than the connected load R_L, the Johnson input noise voltage is given by

$$\left(\overline{v_n^2}\right)_{\text{in}} = 4kT_0 R_L B . \tag{2.30}$$

At the output this noise power is $(\overline{v_n^2})_{\text{out}} = G^2(\overline{v_n^2})_{\text{in}}$. Taking the excess noise voltage from the amplifier, seen at the output, as v_{exc}^2 we write (2.29) as

$$F = \frac{S_{\text{in}}}{S_{\text{out}}} \frac{N_{\text{out}}}{N_{\text{in}}} = \frac{1}{G^2} \frac{G^2 \left(\overline{v_n^2}\right)_{\text{in}} + \overline{v_{\text{exc}}^2}}{\left(\overline{v_n^2}\right)_{\text{in}}} = 1 + \frac{\overline{v_{\text{exc}}^2}}{G^2 \left(\overline{v_n^2}\right)_{\text{in}}} . \tag{2.31}$$

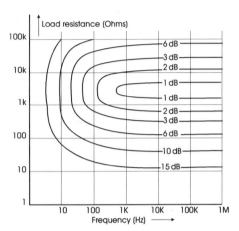

Fig. 2.4. Noise figure contours for a typical low-noise amplifier

Thus F, besides depending on R_L and frequency, is also a function of temperature as is included in (2.30). Assuming the amplifier always kept at the same temperature of 290 K, so that $\overline{v_{exc}^2}$ depends only on R_L and frequency, we derive from (2.31) the value of F_T for a different source temperature but the same frequency and load resistance. We find

$$F_T = 1 + \frac{T_{290}}{T}(F - 1) . \tag{2.32}$$

The effective input noise voltage $(\overline{v_n^2})_{in, \, eff}$ defined by $G^2(\overline{v_n^2})_{in, \, eff} = N_{out}$ is equal to

$$\left(\overline{v_n^2}\right)_{in, \, eff} = \left(\overline{v_n^2}\right)_{in} + \frac{\overline{v_{exc}^2}}{G^2} . \tag{2.33}$$

According to (2.31) $\overline{v_{exc}^2}/G^2 = 4kT_{290}R_L B(F - 1)$. For an arbitrary temperature T we now have $(\overline{v_n^2}) = 4kTR_L B$ so that

$$\left(\overline{v_n^2}\right)_{in, \, eff} = 4kBR_L \left[T + (F - 1)T_{290}\right] . \tag{2.34}$$

Defining an effective temperature T_{eff} by

$$T_{eff} = T + (F - 1)T_{290} \tag{2.35}$$

and comparing (2.34) with (2.30) it is seen that the excess noise of the amplifier can be simply taken into account by calculating the Johnson noise of the source resistance for the effective temperature given by (2.35).

The other noise powers accompanying the input signal which were so far left out can be taken into account, just like the signal power, by amplifying them with the factor G. Thus the signal-to-noise factor at the amplifier output is the same as at the input except for the effective temperature of source impedance.

3

Thermal Detectors

Thermal detection is based on temperature changes when exposed to radiation. The temperature change on its turn induces electrical power or passively it changes the electrical properties of an electronic circuit element. These radiation transducers are relatively simple and cheap and meet the requirements of many applications. Nevertheless the most suitable materials and detector constructions must be selected for reaching high performance of sensitivity and responsivity. For many applications those thermal detectors do better than or are compatible with photon detectors.

The main distinction with photon detectors which are based on the creation of free carriers, is the radiation absorption by the lattice which causes its heating. The change in temperature of the lattice effects the electronic system of the material like thermoelectric power, resistance, or electrical polarization.

Due to the thermal response the detection speed is low and alternating output signals are therefore limited to low frequencies, except for the pyroelectric detector that can operate even beyond megahertz. Usually the thermal detectors are applied to the infrared and far infrared part of the spectrum where, because of the small photon energy, photon detectors do not exist. Since it is the heat and not the photon size that is relevant these detectors are, in principle, applicable in the whole spectrum provided the radiation will be absorbed.

The developments of thermal detectors have let to instruments with sensitivities and accuracies that are set by fundamental noise fluctuations and thereby reach the fundamental detection limit. It is found from the analysis and material choices that a large response and a fast rise time of the output signal are not always compatible with high sensitivity or low minimum detectable power. In the following the operating parameters and ultimate sensitivity of the most common thermal detectors will be analyzed together with the conditions for the best performance.

3.1 Thermocouple and Thermopile

A very simple and common instrument to measure temperatures in a wide range of applications is the thermocouple. This instrument is also very useful to measure radiation, even at low power density. The principle is based on the temperature dependent thermoelectric power that exists between different metals. An instrument is simply made by a closed circuit of two connected metal wires. A current is generated if one junction is at a higher temperature than the other. A schematic drawing of a thermocouple connected to a radiation receiver is shown in Fig. 3.1. The circuit consists of two wires of metal A and one of metal B. The receiver with junction J_1 is heated by the incident radiation and has a temperature increase ΔT relative to the "cold" junction J_2. By blackening the receiver the radiation absorption can be very efficient. A relative temperature rise of one junction results in a relative increment of the electromotive potential across that junction.

The voltage ΔV_{12} indicated in Fig. 3.1 between two pieces of the same metal A, shown in Fig. 3.1, is for small values of ΔT proportional or

$$\Delta V = P_{AB}\Delta T, \tag{3.1}$$

where P_{AB} is the thermoelectric power between metals A and B. The thermal potential of elements have been measured relative to a "standard metal". For obtaining P_{AB} one just has to abstract the standard values of the two metals A and B. If the circuit is closed with a resistance R_L the current will be

$$i = \frac{P_{AB}\Delta T}{R + R_L}, \tag{3.2}$$

where R is the resistance of the thermocouple. The current flow consumes some heat from the receiver which is transported to the gold junction. Saying differently, the current has a cooling effect on the junction J_1, which is called the Peltier effect. The amount of heat power consumption is related to P_{AB} by

$$\Delta W = iTP_{AB}. \tag{3.3}$$

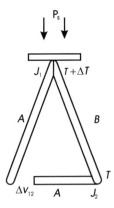

Fig. 3.1. Principle of thermoelectric power between two different metals A and B. If junctions J_1 and J_2 are at different temperatures a voltage difference ΔV_{12} can be observed

Let us now calculate the potential difference ΔV_{12} caused by a constant amount W of radiation power absorbed by the receiver and by simultaneously taking into account the Peltier effect of the generated current. The thermal conductance of the heated junction together with the receiver is λ. We have the following heat equation

$$W - iTP_{AB} = \lambda \Delta T. \tag{3.4}$$

Solving i and ΔT from the (3.2) and (3.4) we get

$$\Delta T = \frac{W}{\lambda} \frac{R + R_{\mathrm{L}}}{R + R_{\mathrm{L}} + R_{\mathrm{d}}}. \tag{3.5}$$

$$i = \frac{W P_{AB}}{\lambda \left(R + R_{\mathrm{L}} + R_{\mathrm{d}}\right)}, \tag{3.6}$$

where

$$R_{\mathrm{d}} = \frac{T P_{AB}^2}{\lambda} \tag{3.7}$$

is the so-called dynamic resistance of the thermocouple. In the absence of the Peltier effect the temperature increase would be higher at $\Delta T_0 = W/\lambda$ so that we may write (3.5) as

$$\Delta T = \Delta T_0 \frac{R + R_{\mathrm{L}}}{R + R_{\mathrm{L}} + R_{\mathrm{d}}}. \tag{3.8}$$

For the usual thermocouples the value of R_{d} is small compared with R. The maximum signal voltage $V_s = iR_{\mathrm{L}}$ is measured with an input impedance R_{L} of the amplifier which is much larger than R. We derive from (3.6) that $V_s = W P_{AB}/\lambda$ is independent of both R and R_{d}. However R and R_{d} are both non-negligible in finding the minimum detectable power of the thermocouple.

For signal processing it is often desirable to have the signal alternating. In practice this is realized by chopping a constant heat flow. To analyze this situation we consider an alternating incident radiation in the form of $W e^{j\omega t}$ with $\omega = 2\pi f$. The temperature and current will then of course fluctuate with the same frequency so we write $\Delta T e^{j\omega t}$ for the temperature and $i e^{j\omega t}$ for the current. The time dependence of the heat balance includes now also the thermal capacity. We then find for the heat equation

$$W - iTP_{AB} = i\omega C_{\mathrm{th}} \Delta T + \lambda \Delta T. \tag{3.9}$$

Combining (3.2) and (3.9) we find

$$\Delta T = \frac{W}{\lambda \left(1 + j\omega \tau_{\mathrm{th}}\right)} \frac{R + R_{\mathrm{L}}}{R + R_{\mathrm{L}} + Z_{\mathrm{d}}} \tag{3.10}$$

$$i = \frac{W P_{AB}}{\lambda \left(1 + j\omega \tau_{\mathrm{th}}\right) \left(R + R_{\mathrm{L}} + Z_{\mathrm{d}}\right)}, \tag{3.11}$$

where $\tau_{\text{th}} = C_{\text{th}}/\lambda$ is the thermal time constant and

$$Z_{\text{d}} = \frac{T P_{AB}^2}{\lambda \left(1 + j\omega\tau_{\text{th}}\right)} \tag{3.12}$$

is the dynamic impedance, which is electrically equivalent with a resistance R_{d} shunted by a dynamic capacitance

$$C_{\text{d}} = \frac{C_{\text{th}}}{T P_{AB}^2}. \tag{3.13}$$

It is seen that

$$\tau_{\text{th}} = \frac{C_{\text{th}}}{\lambda} = R_{\text{d}} C_{\text{d}}. \tag{3.14}$$

The signal voltage is $V_{\text{s}} = iR_{\text{L}}$. The responsivity r is then given by $|V_{\text{s}}|/W$. In practice the maximum signal voltage is measured with an amplifier for which the impedance $R_{\text{L}} \gg |R + Z_{\text{d}}|$ We may then derive from (3.11)

$$r = \frac{|V_{\text{s}}|}{W} = \frac{P_{AB}}{\lambda \left(1 + \omega^2\tau_{\text{th}}^2\right)^{1/2}}. \tag{3.15}$$

The fluctuations that determine its minimum detectable power arise from the Johnson noise of the ohmic resistance of the thermocouple and from the thermal fluctuations of the receiver. The Johnson noise according to (1.9) is

$$\overline{v_{\text{nJ}}^2} = 4kTRB, \tag{3.16}$$

where B is electronic bandwidth of the system. (Because R and R_{L} are for the noise voltage parallel the contribution of R_{L} is negligible.) The thermal fluctuations produce a voltage noise which can be derived from (1.82) and (3.2). With the condition $R_{\text{L}} \gg R$ it becomes

$$\overline{v_{\text{nT}}^2} = \frac{4kT^2 P_{AB}^2 B}{\lambda \left(1 + \omega^2\tau_{\text{th}}^2\right)}. \tag{3.17}$$

The total noise voltage is

$$\overline{v_{\text{t}}^2} = 4kTB \left[R + \frac{T P_{AB}^2}{\lambda \left(1 + \omega^2\tau_{\text{th}}^2\right)}\right]. \tag{3.18}$$

The signal-to-noise value for a thermocouple follows from (3.15) and (3.18). We obtain

$$\frac{S}{N} = \frac{W^2}{4kTB \left[\lambda T + \frac{\lambda^2 R \left(1 + \omega^2\tau_{\text{th}}^2\right)}{P_{AB}^2}\right]}. \tag{3.19}$$

From this we have

$$\text{NEP}^2 = 4kT^2 B\lambda \left[1 + \frac{R}{R_{\text{d}}} \left(1 + \omega^2\tau_{\text{th}}^2\right)\right], \tag{3.20}$$

where we have used (3.7). The first term of (3.20) comes from the thermal fluctuations and the second from the Johnson noise. The ratio of the two contributions for frequencies below that corresponding to the reciprocal thermal time constant is

$$\frac{\overline{v_{\mathrm{nJ}}^2}}{\overline{v_{\mathrm{nT}}^2}} = \frac{\lambda R}{T P_{AB}^2} = \frac{R}{R_{\mathrm{d}}} .$$

(3.21)

For practical thermocouples R_{d} is not more than 10 percent of R so that the NEP is mainly determined by Johnson noise only.

It is seen from (3.15) and (3.20) that the lower the thermal conductivity the higher the responsivity and the lower the NEP. The thermal conductivity is due to the radiation losses of the receiver, convection by the surrounding air and by conduction along the leads of the couple. The convection heat by the surrounding air can be eliminated by evacuating the air. The ultimate minimum of the thermal conductance is set by radiation losses only. It is therefore challenging to reduce the conduction losses along the leads as far as possible and to get it at most comparable with the radiation losses. Leads as thin as $100\,\mu\mathrm{m}$ have been used. On the other hand it is desirable to keep the electrical resistance as low as possible to reduce the NEP as follows from (3.20). Unfortunately these requirements for thermal and electrical conductivity are not compatible, because according to the Wiedemann-Franz law it is stated that metals with low thermal conductivity have also low electrical conductivity. Nevertheless this difficulty can be reduced by optimizing the design parameters.

Choosing an arbitrary (small) value of the thermal conductance λ for obtaining the desired responsivity the question arises how to distribute λ over the two leads so that the electric resistance has a minimum. The answer can be given by minimizing the product λR. This happens when the ratios R/λ for both leads are equal. This can be seen as follows. The total in series resistance R and total in parallel thermal conductivity λ of the two leads are given by

$$R = \frac{\rho_1 l_1}{a_1} + \frac{\rho_2 l_2}{a_2}$$

(3.22)

$$\lambda = \frac{a_1 q_1}{l_1} + \frac{a_2 q_2}{l_2} ,$$

(3.23)

where ρ_1, q_1, l_1, a_1, and ρ_2, q_2, l_2, a_2 are the specific electrical resistivity, the specific thermal conductivity, the length and the cross section for the two leads. The two equations have two variables l_1/a_1 and l_2/a_2. To minimize λR we keep one variable constant and we look for the zero condition of the derivative of λR with respect to the other one. We then find $R_1/\lambda_1 = R_2/\lambda_2$. Thus λ is distributed over the two leads in such a way that the latter condition is fulfilled.

The desired high responsivity with small λ may lead to a long thermal time constant and consequently to a small frequency range. However, it may be possible to maintain a relatively high frequency range by minimizing the heat capacitance which is contributed by the receiver and parts of the leads near the heated junction. Often a gold foil, black on one side to avoid reflections, is used with, for instance, a thickness of $0.3\,\mu$m and an area of $0.5\,\text{mm}^2$.

For the realization of a thermocouple the choice of a metal combination with the most suitable thermal potential has to be made for the ambient temperature of operation. The geometrical design parameters are determined by the technology to miniaturize the couple with receiver. At room temperature the combinations of bismuth with antimony, bismuth with tellurium, and iron with constantan (Cu–Ni alloy) are usual. At low temperature the sensitivity decreases considerably and the best choice is then the combination of a gold alloy having 1% cobalt with copper. At high temperature up to 1700° C the combination of platinum with a platinum alloy containing 15% rhodium is a good choice.

With a number of thermocouples in series to form a thermopile the total thermoelectric power is proportional to the number n. The electrical resistance and the thermal conductivity are also proportional to n so that the responsivity is more or less the same as for a single thermocouple whereas the NEP increases with the square root of n. The thermal time constant is divided by n so that the frequency range may roughly increase with n assuming that the thermal capacitance is mainly determined by the foil of the receiver.

Example

We consider a bismuth–antimony thermocouple with $P_{AB} = 10^{-4}\,[\text{V K}^{-1}]$, $\lambda = 5 \times 10^{-5}\,[\text{W K}^{-1}]$, and $R = 3\,\Omega$. By using (3.15) with $\omega = 0$ we obtain a responsivity of $2\,[\text{V W}^{-1}]$ and by using (3.20) we obtain a NEP of about $1.1 \times 10^{-10}\,[\text{W Hz}^{-1/2}]$.

3.2 Bolometer

The bolometer is a detector element whose electrical resistance is a sensitive function of temperature. The absorbed energy of the incident beam heats the detector element and changes its resistance which is then observed by the voltage change of the bias current through this detector element. Metals, semiconductors, and superconductors are used as sensitive element. Especially superconducting transitions are very sensitive giving a rapid change of resistance in a relatively small temperature domain but they are for that reason limited to specific ambient temperatures.

The elements are usually in the form of a sputtered film or a thin flake in order to have low heat capacitance and therewith relatively fast responsivity. A schematic arrangement of a bolometer is shown in Fig. 3.2. It consists of a temperature sensitive element with resistance R which is in series with a load resistance R_L connected to a battery with voltage V. The side of the sensitive element that is exposed to the radiation is blackened for maximum absorption. The incident radiation changes the resistance on heating. This change is described by a temperature coefficient α defined as

$$\alpha = \frac{dR}{R dT}, \tag{3.24}$$

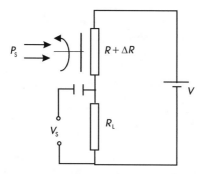

Fig. 3.2. Principle of bolometer. Chopped incident radiation P_s changes the resistance on heating. The induced circuit current is observed as a signal voltage V_s

where R is the resistance at temperature T. Metals have a positive temperature coefficient of about $3 - 4 \times 10^{-3}$ [K^{-1}]. The coefficient does not vary much for most metals and is to a good approximation more or less equal to T^{-1}. For semiconductors α is negative and the absolute values may be an order of magnitude larger. Their values of R are in general a strong function of temperature. Well above the absolute zero temperature a good approximation is given by

$$R = R_0 T^{-3/2} e^{A/T},\qquad(3.25)$$

where R_0 and A are material constants. A typical value for A is $3{,}000$ K. We obtain from the last equation $\alpha = -(A/T^2) - (3/2T)$ which gives at room temperature $\alpha = -3.8 \times 10^{-2}$ [K^{-1}].

Larger values of $|\alpha|$ in the order of 0.5 to 3 [K^{-1}] are obtained with heavily doped semiconductors at cryogenic temperatures. For example gallium doped single crystal germanium at 2 K [9] and phosporous doped silicon below 20 K [36]. The heat capacity of these semiconductors is due to lattice vibrations and falls sharply at low temperature approaching zero as $(T/T_d)^3$ where T_d is the Debye temperature. T_d is 366 K for germanium and 658 K for silicon. Thus at low temperature the specific heat capacity for silicon is at least five times smaller than that of germanium. The smaller heat capacity permits a higher frequency response.

The change in resistance ΔR for small changes of ΔT becomes

$$\Delta R = \alpha R \Delta T.\qquad(3.26)$$

The observed signal V_s is measured over R_L. For signal processing and amplification the input is alternating, mostly by chopping the incident radiation. To eliminate the dc voltage over R_L and undesirable dc radiation signals from the background, the observed signal passes a blocking capacitor as shown in Fig. 3.2. The resistance R_L is chosen much larger than R for maintaining constant current and therewith getting a maximum signal. The maximum ac signal becomes

$$V_s = i_0 \Delta R = \alpha i_0 R \Delta T = \alpha i_0 R \Delta T.\qquad(3.27)$$

It is seen that the signal increases with the bias current so it is of interest to optimize i_0. The question arises about the temperature stability of the heated resistor because a change of resistance changes the ohmic heating which on its turn changes the resistance again. We analyze this as follows. The additional heating of the resistor by the bias current after a change ΔT due to the change ΔR is by using (3.26) given by

$$\Delta W = \alpha i_0^2 R \Delta T. \tag{3.28}$$

Let us consider an ambient temperature T_0 and a homogeneous temperature T of the detector element by the ohmic heating $(i_0^2 R)$. Then the equilibrium temperature is given by the balance between heating and cooling or

$$i_0^2 R = \lambda \left(T - T_0 \right), \tag{3.29}$$

where λ is the thermal conductivity. Next we introduce a small disturbance around T which might be caused by any source. This disturbance δT causes additional ohmic heating and the corresponding heat equation is

$$C_{\text{th}} \frac{\mathrm{d}T}{\mathrm{d}t} = \alpha i_0^2 R \delta T - \lambda \delta T, \tag{3.30}$$

where C_{th} is the heat capacity. If the right hand side of the last equation is negative we have stability. The stability condition is then written as

$$\lambda > \alpha i_0^2 R. \tag{3.31}$$

As mentioned the α-values for metals can be approximated by $1/T$. We then find by substituting this relation for α in (3.31) and using (3.29) that stability is obtained when

$$T > T - T_0 \tag{3.32}$$

which always holds. Also for semiconductors where $\alpha < 0$ we have stability.

Next we consider alternating incident radiation in the form of $W e^{j\omega t}$ with $\omega = 2\pi f$. The temperature and resistance will then of course fluctuate with the same frequency so we write $\Delta T e^{j\omega t}$ for the temperature change and similar for the change of resistance. Assuming $|\Delta R| \ll R$ the heat equation becomes

$$j\omega C_{\text{th}} \Delta T + \lambda \Delta T = W + \alpha i_0^2 R \Delta T. \tag{3.33}$$

Solving for ΔT we get

$$\Delta T = \frac{W}{\lambda_e \left(1 + j\omega \tau_e \right)}, \tag{3.34}$$

where

$$\lambda_e = \lambda - \alpha i_0^2 R \tag{3.35}$$

is the effective conductivity and

$$\tau_e = \frac{C_{\text{th}}}{\lambda_e} \tag{3.36}$$

is the effective time constant. Next we substitute (3.27) into (3.34) and find for the response

$$r = \frac{V_s}{W} = \frac{\alpha i_0 R}{\lambda_e \left(1 + j\omega\tau_e\right)} \,. \tag{3.37}$$

It is seen that for high response λ_e should be as small as possible. For applying relatively high frequency τ_e should be as small as possible, which means that requiring already small λ_e we have to reduce C_{th} as far as possible. It looks at first glance also advantageous to increase i_0 as far as possible for higher response. The optimum value of i_0 is, however, often limited by the desired minimum NEP or sensitivity.

3.2.1 Metallic Bolometer

By considering the noise we assume a uniform temperature T for the detector element and an ambient temperature T_0. The thermal power fluctuations are given by (1.82). It was pointed out in sect. 1.7.1 that in the case of radiation one half is due to fluctuating emission and the other half due to fluctuating absorption. In the case of heat transfer by conduction the fluctuations are also equally contributed by the heat flows in two directions so that one half of the noise is determined by the body temperature T whereas the other half by the surrounding at T_0. The voltage fluctuations at frequency ω within the bandwidth B are given by

$$\overline{v_{nT}^2} = \left(2kT_0^2 + 2kT^2\right) \frac{\left(\alpha i_0 R\right)^2 B}{\lambda_e \left(1 + \omega^2 \tau_e^2\right)} \,, \tag{3.38}$$

where we have used (3.27). Because of the electrothermal feedback by the bias current the effective conductivity λ_e and effective time constant τ_e instead of the real physical conductivity λ and time constant τ are found in (3.38). The Johnson noise for $R_L \gg R$ is according to (1.9) given by

$$\overline{v_{nJ}^2} = 4kTRB \,. \tag{3.39}$$

The amplification of the Johnson noise by the electrothermal feedback can be neglected because of the small $|\alpha|$-value. Further, although at low frequencies, say below $10\,\mathrm{Hz}$, some current noise may appear when very thin wires or films are used, it is generally negligible. The total noise voltage becomes

$$\overline{v_t^2} = \overline{v_{nT}^2} + \overline{v_{nJ}^2} \,. \tag{3.40}$$

The signal-to-noise ratio for a metallic bolometer follows from (3.37) and (3.40). Using for metal $\alpha = 1/T$ and substituting (3.29) into (3.35) we get $\lambda_e = \lambda T_0/T$. We then obtain

$$\frac{S}{N} = \frac{W^2}{T_0^2 \lambda B \left[2k\frac{T_0}{T} + 2k\frac{T}{T_0} + \frac{4kT}{T-T_0}\left(1 + \omega^2\tau_e^2\right)\right]} \,. \tag{3.41}$$

From this we obtain the NEP at zero frequency as

$$\text{NEP}^2 = T_0^2 \lambda B \left[2k\frac{T_0}{T} + 2k\frac{T}{T_0} + \frac{4kT}{T - T_0} \left(1 + \omega^2 \tau_e^2\right) \right]. \tag{3.42}$$

The term $(T_0/T) + (T/T_0)$ within the brackets associated with the thermal noise indicates an increasing NEP with temperature, whereas the third term associated with the Johnson noise decreases with temperature. Starting from room temperature $T = T_0$ and $\omega = 0$ the NEP decreases with T and reaches a minimum for $T = 2.54T_0$. This high temperature is in general not realistic in practice. At a more practical temperature of $1.5T_0$ the NEP is about 10% above its minimum.

Metals mostly used as sensitive elements are gold, platinum, and nickel which has the highest α-value. Because metals have high reflectivity, especially for infrared radiation they are black coated and made of thin evaporated layers on non–conducting substrates. As is seen from (3.37) the thermal conductivity must be low for high response. The highest value is then obtained by radiation cooling only. For having both high and rapid response (small τ_e) the heat capacity is minimized. Thin evaporated metal films of nickel have been made with a thickness of less than $0.1\,\mu$m and a time constant of only $4\,$ms [5, 6].

Example

We consider for a metallic bolometer with $R = 100\,\Omega$, $T_0 = 300\,$K, $T = 500\,$K, $C_{th} = 3 \times 10^{-7}\,\left[\text{J K}^{-1}\right]$, $f = 20\,$Hz, $\alpha = 2 \times 10^{-3}\,\left[\text{K}^{-1}\right]$, and $\lambda = 10^{-4}\,\left[\text{W K}^{-1}\right]$. We obtain with (3.29), (3.35–3.37) and (3.42), respectively, $i_0 = 1.4 \times 10^{-2}\,[\text{A}]$, $\lambda_e = 0.6 \times 10^{-4}[\text{W K}^{-1}]$, $\tau_e = 5\,$ms, $|r| = 47\,[\text{V W}^{-1}]$, and NEP$= 4.7 \times 10^{-11}\,\left[\text{W Hz}^{-1/2}\right]$.

3.2.2 Thermistor

Semiconductor bolometers are highly developed detectors for low power beams and are especially applied in the infrared and submillimeter spectral ranges. For obtaining high responsivity and sensitivity (low NEP) these detectors are operated at very low temperature. If the active element of the bolometer is a thin semiconductor film usually composed of oxidic mixtures of manganese, nickel, and cobalt the name thermistor (thermally sensitive resistor) is often used. These materials, usually applied at room temperature, are good radiation absorbers, although some blackening may be attached to the surface to avoid reflection. The thermistor has a high negative temperature coefficient which even at room temperature provides high responsivity in the order of $700 - 1200\,[\text{V W}^{-1}]$ [8].

The instrumental parameters like r, τ_e, and λ can be obtained experimentally by measuring the voltage–current curve. A typical curve for a semiconductor element (negative α-value) is shown in Fig. 3.3. This curve refers to

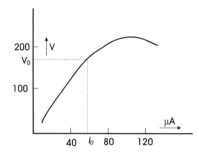

Fig. 3.3. Voltage–current curve of a thermistor

a square flake of thermistor material containing a mixture of the oxides of manganese, nickel, and cobalt having a resistivity of 250 $[\Omega\,\text{cm}]$. The area of the flake is $1\,\text{mm}^2$ and its thickness $10\,\mu\text{m}$. The ohmic heating decreases the resistance and the temperature is determined by the conductivity as given by (3.29). For the operating current i_0 the voltage V_0 is indicated. The resistance of the element is $R = V/i$ but its deviation from the derivative $Z = \mathrm{d}V/\mathrm{d}i$ contains information on λ and α. As a consequence the responsivity can be described by the voltage–current curve. For this purpose we describe $\mathrm{d}V/\mathrm{d}i$ in terms of the dissipated electrical power P by substituting $V = (PR)^{1/2}$ and $i = (P/R)^{1/2}$. We write

$$Z = \frac{\mathrm{d}V}{\mathrm{d}i} = \frac{1/2\,(PR)^{-1/2}\left[R\frac{\mathrm{d}P}{\mathrm{d}R} + P\right]}{1/2\left(\frac{P}{R}\right)^{-1/2}\left[\frac{1}{R}\frac{\mathrm{d}P}{\mathrm{d}R} - \frac{P}{R^2}\right]} = R\frac{\frac{\mathrm{d}P}{\mathrm{d}R} + \frac{P}{R}}{\frac{\mathrm{d}P}{\mathrm{d}R} - \frac{P}{R}}. \tag{3.43}$$

By using (3.24) and (3.29) we derive

$$\frac{\mathrm{d}P}{\mathrm{d}R} = \frac{\mathrm{d}P}{\mathrm{d}T}\frac{\mathrm{d}T}{\mathrm{d}R} = \frac{\lambda}{\alpha R}. \tag{3.44}$$

Next we substitute (3.44) into (3.43) to obtain

$$\frac{\lambda}{\alpha Ri^2} = \frac{\lambda}{\alpha P} = \frac{Z + R}{Z - R}. \tag{3.45}$$

Substituting (3.45) into (3.37) and using (3.35) a useful relation for the responsivity is obtained [10]

$$r = \frac{1}{2i}\left(\frac{Z - R}{R}\right)\frac{1}{1 + \mathrm{j}\omega\tau_{\mathrm{e}}}. \tag{3.46}$$

We note that, as expected, for $Z = R$, which means that the voltage–current line is straight, the responsivity is zero. The time constant τ_{e} can also be expressed in terms of the voltage–current curve. Using (3.35) and (3.45) we find

$$\tau_{\mathrm{e}} = \frac{C_{\mathrm{th}}}{\lambda - \alpha P} = \tau\frac{Z + R}{2R}. \tag{3.47}$$

In practice the temperature coefficient α of the resistance of the detection element is known. Then using this parameter the heat conductivity λ of the element can be obtained experimentally by means of (3.45) from the voltage–current curve. Measuring the responsivity r as a function of frequency the time constant τ_e is obtained and with the aid of (3.47) the heat capacity C_{th}.

Calculating the noise equivalent power from the noise the electrothermal feedback from the bias source must be taken into account which in the case of a semiconductor can be considerably [7]. We consider as before the load resistance $R_L \gg R$ so that i is constant. When the Johnson noise voltage is added to the bias voltage over the detection element the power P dissipated by the constant current i increases and because of the negative α the resistance R decreases. This effect will reduce the voltage over the detector and thus reduce the observable noise voltage. Similarly when the noise voltage is opposed and subtracted from the bias voltage the noise voltage decreases the power dissipation, decreases the temperature and increases the resistivity. The voltage across the detector increases and thus the observed noise voltage is again reduced. The Johnson noise voltage according to (1.9) is

$$\left(\overline{v_n^2}\right)^{1/2} = (4kTRB)^{1/2} . \tag{3.48}$$

The power induced by the Johnson noise voltage becomes

$$P_J = i \, (4kTRB)^{1/2} . \tag{3.49}$$

Substituting (3.49) into (3.46) we obtain the feedback voltage $\overline{(v_n^2)}_{fb}^{1/2}$ given by

$$\overline{(v_n^2)}_{fb}^{1/2} = (4kTRB)^{1/2} \left(\frac{Z-R}{2R}\right) \frac{1}{1+j\omega\tau_e} . \tag{3.50}$$

The observed Johnson noise voltage is then the sum of (3.48) and (3.50) or

$$\overline{(v_n^2)}_J^{1/2} = (4kTRB)^{1/2} \left[1 + \frac{Z-R}{2R} \frac{1}{1+j\omega\tau_e}\right] . \tag{3.51}$$

For $Z = 0$ and $\omega = 0$ we have $\overline{(v_n^2)}_J = kTRB$. Thus due to the electrothermal feedback the Johnson noise power is reduced by a factor four.

The signal-to-noise ratio in the case of only Johnson noise is obtained by (3.51) and (3.46) or

$$\left(\frac{S}{N}\right)_J = \frac{v_s^2}{\overline{(v_n^2)}_J} = \frac{\left[\frac{W}{i}\left(\frac{Z-R}{2R}\right)\right]^2}{4kTRB} \left|\frac{\frac{1}{1+j\omega\tau_e}}{1 + \frac{Z-R}{2R}\frac{1}{1+j\omega\tau_e}}\right|^2 . \tag{3.52}$$

The Johnson-limited NEP_J is

$$NEP_J = (4kTPB)^{1/2} \left|\frac{Z+R}{Z-R}\right| \left(1 + \omega^2\tau^2\right)^{1/2} , \tag{3.53}$$

where we have used (3.47). We note that the NEP$_J$ depends on the real physical time constant $\tau = C_{\text{th}}/\lambda$ rather than the effective time constant τ_e. Without electrothermal feedback the NEP$_J$ is for $\omega = 0$ and $Z = 0$ equal to $2(4kTPB)^{1/2}$, a factor two larger.

The derived thermal voltage fluctuations given by (3.38) include the electrothermal feedback by the effective time constant. Calculating the corresponding thermal-noise limited NEP$_T$ with the aid of (3.37) and (3.38) we find

$$\text{NEP}_T = \left[\left(4kT^2 + 4kT_0^2 \right) \frac{\lambda RB}{Z + R} \right]^{1/2} , \tag{3.54}$$

where we have used (3.35) and (3.45). We note that the NEP$_T$ is frequency independent. The total NEP is given by the quadratic combination of the NEP$_s$ from the various sources or

$$\text{NEP} = \left[\sum_i \text{NEP}_i^2 \right]^{1/2} . \tag{3.55}$$

Example

Consider a square flake of sintered semiconductor material $1 \times 1\,\text{mm}^2$ of $10\,\mu\text{m}$ thickness containing a mixture of the oxides manganese, nickel, and cobalt [8]. In practice the applied voltage across the detector element is close to the maximum voltage for which $Z = 0$. The parameters are: $T_0 = 300\,\text{K}$, $\tau = 5 \times 10^{-3}\,\text{s}$, $\lambda = 9 \times 10^{-4}\,[\text{W K}^{-1}]$. Near $Z = 0$ the current $i = 6 \times 10^{-5}\,[\text{A}]$ and the voltage $V = 150\,[\text{V}]$. To avoid the $1/f$ noise the detector is operated at $20\,\text{Hz}$. We calculate $P = 9\,\text{mW}$, $R = 2.5 \times 10^6\,\Omega$, with (3.29) $T = 310\,\text{K}$, with (3.47) $\tau_e = \tau/2$ and with (3.35) and (3.45) $\lambda_e = 2\lambda$. Applying (3.53) we obtain NEP$_J = 1.24 \times 10^{-11}\,[\text{W Hz}^{-1/2}]$ and with (3.54) NEP$_T = 9.6 \times 10^{-11}\,[\text{W Hz}^{-1/2}]$ so that with (3.55) we have NEP$= 9.7 \times 10^{-11}\,[\text{W Hz}^{-1/2}]$. Using (2.2) we get $D^* = 10^9\,[\text{W}^{-1}\,\text{cm Hz}^{1/2}]$. The responsivity with (3.46) gives $|r| = 8,300\,[\text{V W}^{-1}]$.

For most applications the detector should have a fast response. This is achieved by providing a good conducting thermal sink with high electrical resistance which is cemented to the semiconductor. Usual thermal sinks are quartz, glass, or a very thin air gap between detector and a metal thermal conductor. The larger the thermal conductivity the faster the responsivity. With quartz an effective time from 2 to $5\,\text{ms}$, with glass from 5 to $8\,\text{ms}$ and with an air gap from 20 to $50\,\text{ms}$ was obtained [8]. For the chosen material the thickness of the square detector element determines the resistivity. A high resistivity of above $1\,\text{M}\Omega$ is desirable to have the amplifier noise less than the Johnson noise . It is seen from (3.53) and (3.54) that cooling is effective in reducing the NEP. This advantage has led to the development of cryogenic bolometers.

Example

Consider a gallium-doped single crystal germanium in the liquid helium temperature range [9]. A detector element of this material has a sensitive area of $0.15\,\mathrm{cm^2}$, a temperature of $2.15\,\mathrm{K}$, a resistivity of $1.2 \times 10^4\,\Omega$, a bias current of $6.5 \times 10^{-5}\,\mathrm{A}$ and a heat conductivity of $1.83 \times 10^{-4}\,[\mathrm{W\,K^{-1}}]$. The effective time constant is $4 \times 10^{-4}\,\mathrm{s}$ and the operating frequency $200\,\mathrm{Hz}$. The noise from the bias current and flikker noise of this p-type germanium is for the applied bias current at frequencies above $20\,\mathrm{Hz}$ negligible. The voltage across the element is $iR = 0.8\,\mathrm{V}$. The power $P = i^2R = 5 \times 10^{-5}\,\mathrm{W}$ and the temperature $T = P/\lambda + T_0 = 2.42\,\mathrm{K}$ Assuming operation around $Z = 0$ of the power curve we obtain with (3.46) $r = 8,000\,[\mathrm{V\,W^{-1}}]$. With (3.53) and (3.54) we find $\mathrm{NEP_J} = 9.7 \times 10^{-14}$ and $\mathrm{NEP_T} = 4.6 \times 10^{-13}$. With (3.55) we get $\mathrm{NEP} = 4.7 \times 10^{-13}\,[\mathrm{W\,Hz^{-1/2}}]$

In the above example we substituted for the thermal noise $T = 2.42\,\mathrm{K}$ and $T_0 = 2.15\,\mathrm{K}$. However, some background radiation at room temperature reaches the detector element through the input aperture. The average value of this background radiation is eliminated by operating the detector in the ac mode and by installing a blocking capacitor. The fluctuating part of it passes the capacitor and mixes with the signal. The field of view of the cooled detector is generally restricted by an aperture which is also kept at low temperature and shields the detector from the outside thermal radiation. As pointed out in Sect. 2.2 the background radiation fluctuations that reach the detector element are then given by

$$\overline{\Delta P^2} = 8\sin^2\left(\frac{\theta_d}{2}\right) AB\sigma k T^5 , \qquad (3.56)$$

where T is the background temperature and θ_d the cone angle of the aperture. Taking $\sin(\theta_d/2) = 0.5$ we find for the background $\mathrm{NEP_B} = 7.5 \times 10^{-12}$ $[\mathrm{W\,Hz^{-1/2}}]$. Comparing this with the Johnson and thermal noise powers we conclude that this system is background limited. However, so far we have considered the full spectrum of the background. If a narrow band pass filter is applied that transmits only radiation within the band width of the signal beam most of the background noise is eliminated and the NEP will be reduced considerably.

3.3 Pyroelectric Detector

A high frequency thermal detector, even up to megahertz and beyond, can be realized on the basis of ferroelectric materials. Those materials are asymmetric crystals that have permanent internal electric dipole moments with strong temperature dependence. Under equilibrium conditions the electrical asymmetry of the polarization is compensated by free charges on the end surfaces perpendicular to the polarization direction of a suitable sample. An increase in temperature caused by incident radiation has a decremental effect on the polarization which decreases the compensated surface charges. If the temperature change is faster than the process during which these compensated charges redistribute themselves, a potential difference across the material is observed. Thus, these devices are inherently ac detectors At thermal heating frequencies above that corresponding to the reciprocal thermal time constant, usually about 20 Hz, the responsivity, as will be shown, is constant and it leads to a good high frequency performance and its sensitivity can be higher than any other uncooled thermal detector. In principal there is no instrumental deformation of an observed pulsed input signal having a broad frequency spectrum.

A pyroelectric element can be considered as a capacitor with a temperature dependent charge. An equivalent circuit diagram is shown in Fig. 3.4. The detector is mostly connected with an operational amplifier having high input impedance (see Sect. 7.1). Successful developments have been obtained with triglycine sulphate, strontium barium niobate, and lithium sulphate. The choice of detector material is determined by a large pyroelectric coefficient, small real and imaginary component of the dielectric constant, and a small thermal capacity. Although the thermal conductivity with the surroundings of the element is not relevant for the responsivity, the conductivity of the material itself is of importance for the homogeneity of the heating.

The change of the surface charge ΔQ of a ferroelectric material due to a small change ΔT of the temperature is given by

$$\Delta Q = A \left(\frac{\mathrm{d}P}{\mathrm{d}T} \right) \Delta T = AK_\mathrm{p}\Delta T \,, \tag{3.57}$$

where the pyroelectric coefficient $K_\mathrm{p} = \mathrm{d}P/\mathrm{d}T$ is the change of polarization with temperature and A is the surface area of the detector element. The current of the detector is the rate of change of charge

$$i_\mathrm{s} = AK_\mathrm{p}\frac{\mathrm{d}T}{\mathrm{d}t} \,. \tag{3.58}$$

The heat equation of the detector element when heated by the incident radiation power P is

$$C_\mathrm{th}\frac{\mathrm{d}T}{\mathrm{d}t} + \lambda \left(T - T_0\right) = P \,, \tag{3.59}$$

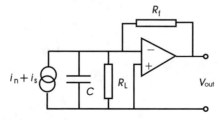

Fig. 3.4. Circuit of pyroelectric detector with operational amplifier

where C_{th} is the thermal capacity, λ the thermal conductance of the element with the surroundings, and T_0 the ambient temperature. To study the frequency response we consider an input power containing a periodic component $P(1 + e^{j\omega t})$. The temperature is then described by

$$T = T_0 + T^* + T(\omega) e^{j\omega t}, \tag{3.60}$$

where $T(\omega)$ is the amplitude of the oscillating temperature component and T^* is the steady state temperature increase by P. Substituting these functions for P and T into (3.59) and considering the steady state we have

$$j\omega C_{th} T(\omega) + \lambda T(\omega) = P \tag{3.61}$$

and from this

$$|T(\omega)| = \frac{P}{\lambda (1 + \omega^2 \tau_{th}^2)^{1/2}}, \tag{3.62}$$

where $\tau_{th} = C_{th}/\lambda$ is the thermal time constant. Substituting (3.60), (3.62) into (3.58) we get for the signal current with frequency ω

$$|i_s| = \frac{\omega A K_p P}{\lambda (1 + \omega^2 \tau_{th}^2)^{1/2}}, \tag{3.63}$$

and for the signal voltage

$$|v_s| = \frac{|i_s| R_L}{\left[1 + (\omega R_L C)^2\right]^{1/2}}, \tag{3.64}$$

where C is the electrical capacitance and R_L the load or shunt resistance which is usually much smaller than the internal resistance of the detector.

The response given by $|r| = |v_s|/P$ becomes by using (3.63) and (3.64)

$$|r| = \frac{\omega A K_p}{\lambda (1 + \omega^2 \tau_{th}^2)^{1/2}} \frac{R_L}{\left[1 + (\omega R_L C)^2\right]^{1/2}}. \tag{3.65}$$

It is seen that for low frequencies $\omega < 1/\tau_{th}$ the response becomes

$$r = \frac{\omega A K_p R_L}{\lambda} \tag{3.66}$$

and for high frequencies, $1/\tau_{th} < \omega < 1/R_L C$, the response is constant and given by

$$r = \frac{A K_p R_L}{C_{th}}. \tag{3.67}$$

The smaller the load resistance the larger the frequency range with constant response which is independent of both frequency and thermal conductivity.

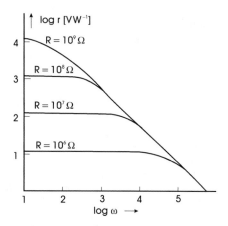

Fig. 3.5. Response for various values of load resistance. $K_p = 2 \times 10^{-4}$ [C m^{-2} K^{-1}]; $C_{th} = 1.64 \times 10^{-5}$ [JK^{-1}]; $C = 22$ pF; $A = 1$ mm^2

A plot of the response as a function of frequency is shown in Fig. 3.5 where we have used the parameters quoted in the example later.

The noise is produced by the resistance (Johnson noise mainly from the shunt resistance) and by the thermal fluctuations. The thermal noise amplitude per unit frequency at frequency ω is $\Delta T(\omega)$. Its derivative $\mathrm{d}/\mathrm{d}t(\Delta T(\omega))$ can be taken as $\omega \Delta T(\omega)$. The corresponding noise current is then $i_n(\omega) = \omega A K_p \Delta T$. Then the mean square current fluctuations is

$$\overline{i_n^2}(\omega) = \omega^2 A^2 K_p^2 \overline{\Delta T^2}(\omega) . \tag{3.68}$$

Substituting (1.82) into (3.68) we get for the thermal noise within the bandwidth B

$$\overline{i_{nT}^2} = \frac{4k\omega^2 A^2 K_p^2 T^2 B}{\lambda \left(1 + \omega^2 \tau_{th}^2\right)} .$$

The total noise current including Johnson noise becomes

$$\overline{i_n^2} = \frac{4kTB}{R_L} + \frac{4k\omega^2 A^2 K_p^2 T^2 B}{\lambda \left(1 + \omega^2 \tau_{th}^2\right)} . \tag{3.69}$$

For the signal-to-noise ratio using (3.63) we find

$$\frac{S}{N} = \frac{\left|i_s^2\right|}{\overline{i_n^2}} = \frac{P^2}{\frac{4kT\lambda^2 \left(1 + \omega^2 \tau_{th}^2\right) B}{\omega^2 A^2 K_p^2 R_L} + 4kT^2 \lambda B} . \tag{3.70}$$

It turns out that the Johnson noise is much larger than the thermal noise so we may write

$$\mathrm{NEP}^2 = \frac{4kT\lambda^2 \left(1 + \omega^2 \tau_{th}^2\right) B}{\omega^2 A^2 K_p^2 R_L} . \tag{3.71}$$

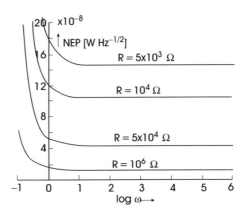

Fig. 3.6. Noise equivalent power for various values of load resistance. $K_p = 2 \times 10^{-4}\,[\mathrm{C\,m^{-2}\,K^{-1}}]$; $C_{\mathrm{th}} = 1.64 \times 10^{-5}\,[\mathrm{J\,K^{-1}}]$; $A = 1\,\mathrm{mm^2}$; $T = 300\,\mathrm{K}$

For frequencies above the value corresponding to the reciprocal thermal time constant we have

$$\mathrm{NEP}^2 = \frac{4kTC_{\mathrm{th}}^2 B}{A^2 K_p^2 R_{\mathrm{L}}}. \tag{3.72}$$

The NEP is plotted in Fig. 3.6 for different values of R_{L} and the parameters used in the following example.

Example

We consider triglycine sulfate with $K_p = 2 \times 10^{-4}\,[\mathrm{C\,m^{-2}\,K^{-1}}]$, specific thermal capacitance $C_s = 1.64 \times 10^6\,[\mathrm{J\,m^{-3}\,K^{-1}}]$, $\lambda = 10^{-5}\,[\mathrm{W\,K^{-1}}]$, and $\varepsilon_r = 25$. Taking a detector area $A = 1\,\mathrm{mm^2}$ and a thickness $t = 10\,\mu\mathrm{m}$ the electrical capacitance $C = \varepsilon_r\varepsilon_0 A/t = 22\,\mathrm{pF}$. An electrical cut off frequency of $10\,\mathrm{kHz}$ requires a load resistance of $R = 1/2\pi fC = 7.2 \times 10^5\,\Omega$. The thermal capacity $C_{\mathrm{th}} = C_s At = 1.64 \times 10^{-5}\,[\mathrm{J\,K^{-1}}]$. With (3.67) we find $r = 8.8\,[\mathrm{V\,W^{-1}}]$. Further with the aid of (3.72) we find $\mathrm{NEP} = 1.2 \times 10^{-8}\,[\mathrm{W\,Hz^{-1/2}}]$. Applying (2.2) we find $D^* = 8 \times 10^6\,[\mathrm{W^{-1}\,cm\,Hz^{1/2}}]$. For a smaller frequency range the shunt resistance becomes larger and consequently also D^*. The other way by choosing a large bandwidth, for instance $1\,\mathrm{GHz}$, equivalent to a nanosecond rise time, we find $R = 7.2\,\Omega$ and the NEP becomes $4 \times 10^{-6}\,[\mathrm{W\,Hz^{-1/2}}]$. Detecting $10\,\mathrm{ns}$ pulses with a bandwidth of $1\,\mathrm{GHz}$ we are then dealing with a NEP of about $0.12\,\mathrm{W}$ and a noise equivalent pulse energy of $1.2 \times 10^{-9}\,\mathrm{J}$. The damage threshold of thermal power will determine the maximum pulse rate that can be detected, wheras ablation of the detector surface may limit the maximum allowable pulse power.

As is seen from (3.67) the responsivity is inversely proportional to the heat capacitance. Usually the thickness of the pyroelectric element is not more than a few microns, determined by the absorption length of the incident radiation. On the other hand the small heat capacitance gives rise to a relatively large electrical capacitance. This in turn limits the frequency range. A larger frequency range will then require a smaller shunt resistance which decreases the responsivity. This unfavorable effect can be avoided to a large extend by applying an operational amplifier as will be discussed in Sect. 7.1. The amplifier output voltage is equal to $i_s R_f$, where R_f is the feedback resistance of the amplifier. In spite of selecting a small value for R_L at high frequency response the output signal voltage of the amplifier is high and remains independent on the value of R_L. In this way a large frequency range with constant high response can be obtained. Usually the bandwidth of the amplifier exceeds 100 kHz or even 1 MHz so that weak signals at high frequencies are observable with pyroelectric detectors.

It should be noted that although the NEP remains also independent of frequency above the value that corresponds to the reciprocal thermal time constant it depends on R_L because the noise current that accompanies the input signal to the amplifier increases with decreasing R_L. Thus for having constant responsivity over a large frequency range the value of R_L is considerably reduced at the expense of the NEP.

Apart from the pyroelectric effect the choice of the pyroelectric material is also determined by its suitability to design the associated low noise amplifier for high frequency operation. The noise produced by both detector and amplifier is minimized by using the device in the voltage mode with a high input impedance. In practice often a FET is used for high input impedance. The noise contribution of the operational amplifier at high frequency increases with frequency and therewith the NEP of the total system. This effect depends on the shunt resistance. The smaller the shunt resistance the higher the frequencies for which the noise contribution of the amplifier is observable.

Strontium barium niobate is because of its high dielectric constant with its relatively low turn over frequency most suitable at low frequencies whereas lithium sulphate with small dielectric constant is more suitable in the high frequency range. Triglycine sulphate has the disadvantage of a low Curie point at 49°C above which the pyroelectric effect no longer exists. This limits the power load on the element.

4

Vacuum Photodetectors

Vacuum photodetectors, capable of very high time resolution with large voltage responsivity, are based on the photoelectric effect. The physical process is the emission of an electron after the absorption of a photon. This happens in a vacuum tube containing electrodes when a photon falls upon the cathode. An electron will then be emitted provided the photon energy is higher than the absorption energy and the minimum energy to escape into the vacuum. For most cathodes the required photon energy is in the visible and shorter wavelength region of the spectrum. Some special multilayer cathode have also been developed to operate in the infrared. The choice of the cathode material is mainly determined by the incident photon energy. Metals have relatively high reflectivity and large escape energies, indicated by the work function. For most metals the work functions are in the range of 4–5 eV so that the radiation wavelength must be at least smaller than 0.3 µm. They are used for the detection of UV and VUV radiation.

Much lower escape energies are obtained for semiconductors. The choice is determined by the photon energy and quantum efficiency, which is defined as the net efficiency with which the incident photons are converted to emitted electrons. It depends on the photon energy, the effective diffusion length of electrons within the photocathode material and the work function of the surface [14].

A schematic energy diagram of a typical semiconductor photocathode is shown in Fig. 4.1 where the band gap energy E_g and electron affinity χ are indicated. The electron affinity is the difference between the minimum escape energy (vacuum level) and the bottom energy of the conduction band. Although strictly speaking the difference between the vacuum level and the Fermi level is the minimum energy to escape, the Fermi level contains at room temperature and below few electrons so that this minimum energy is ineffective. The required minimum photon energy is usually taken as $W = E_g + \chi$. The energy W may then be considered as the work function of the semiconductor. By developing multilayer photocathodes zero and negative electron affinities have been obtained which result in higher quantum efficiencies and

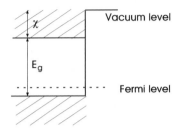

Fig. 4.1. Energy diagram of semiconductor

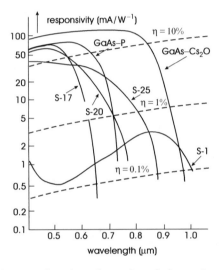

Fig. 4.2. Responsivity as a function of wavelength for various high efficiency photocathodes

lower minimum photon energies [15]. An example is the GaAs photoemittor coated with a thin layer of Cs_2O which has a negative electron affinity of 0.55 eV. This type of detectors can have quantum efficiencies above 10% in the infrared [16]. For several high efficient photocathodes current responsivity and quantum efficiency as a function of radiation wavelength are shown in Fig. 4.2.

4.1 Vacuum Photodiode

The vacuum photodiode consists of a vacuum tube containing a photocathode and a positively biased anode as shown in Fig. 4.3. After the photon absorption the released electron is accelerated toward the anode and driven through the circuit by the applied electric field maintained by a high voltage supply in the range of 100–1,000 V. The resulting current passes the resistance R_L and the signal voltage is measured over this resistor.

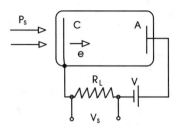

Fig. 4.3. Vacuum photodiode

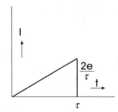

Fig. 4.4. Microcurrent pulse of an accelerated electron in a vacuum photodiode

The transit time of the electron from cathode to anode determines the time resolution or frequency response. The smaller the transit time the larger the frequency range. For high frequency response the applied voltage should be high and the distance between cathode and anode small. Operating at high frequency as determined by the transit time the signal resistance R_L must be small enough so that the diode capacitance does not limit this frequency response. Because of the inherent low diode capacitance it turns out that the resistance for high frequency performance can be high, in the order of several kΩ, so that this type of detector is capable of high frequency response with large signal voltage and relatively low amplifier (or Johnson) noise. They reach frequencies up to GHz with low NEP as compared with semiconductor photodiodes.

To analyze the device quantitatively we consider a plane parallel anode–cathode configuration with an emitting electron travelling from the cathode to the anode as shown in Fig. 4.4. The applied voltage V between cathode and anode gives a constant accelerating field $E = V/d$ to the electron where d is the distance between cathode and anode. The acceleration $a = eE/m$ gives the electron a transit time τ equal to

$$\tau = d\sqrt{\frac{2m}{eV}}. \tag{4.1}$$

Because the electron velocity, $v = at$, increases linearly with time, so does the current. The current of the electron in the external circuit, see appendix A, is then given by

$$i = \frac{2e}{\tau^2}t. \tag{4.2}$$

The Fourier transform of this current pulse according to (1.16) becomes

$$i(\omega) = \frac{2e}{(\omega\tau)^2}\left[(1+j\omega\tau)\,e^{-j\omega\tau} - 1\right].\tag{4.3}$$

The current pulse being the response of the incident photon starts immediately after the photon absorption. The photon input pulse, compared with the current pulse, can be considered as a δ-function for which the Fourier transform has constant amplitudes for all frequency components. The frequency response of this δ-function gives apart from the factor e the frequency response $F(\omega)$ of the detector or

$$F(\omega) = \frac{2}{(\omega\tau)^2}\left[(1+j\omega\tau)\,e^{-j\omega\tau} - 1\right].\tag{4.4}$$

The power response, equal to the square of the modulus of (4.4), is given by

$$|F(\omega)|^2 = \frac{4}{(\omega\tau)^4}\left[4\sin^2\left(\frac{\omega\tau}{2}\right) + (\omega\tau)^2 - 2\omega\tau\sin\omega\tau\right]\tag{4.5}$$

and plotted in Fig. 4.5. The half power width is close to $\omega\tau = \pi$ so that the effective cutoff frequency becomes

$$f_c = \frac{\omega_c}{2\pi} = \frac{1}{2\tau}.\tag{4.6}$$

Using eq. 4.1 we have

$$f_c = \frac{1}{2d}\sqrt{\frac{eV}{2m}}.\tag{4.7}$$

The noise of the electrons produced in a vacuum diode has been treated in Sect. 1.3. The spectral power density of the noise is given by (1.34) and can be considered as shot noise described by (1.29).

Apart from photoemission noise current is also delivered by the so called dark current, due to thermionic emission of the photocathode, which is as a dc current also present in the absence of illumination. A prediction of the order of magnitude of the dark current i_d per unit area is given by the Richardson–Dushman equation which states

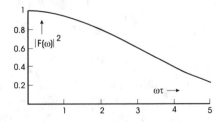

Fig. 4.5. Frequency response of vacuum photodiode

$$i_\mathrm{d} = \frac{4\pi e m_\mathrm{e}}{h^3} \, (kT)^2 \, \mathrm{e}^{-W/kT} \qquad (4.8)$$

or

$$i_\mathrm{d} = 120 T^2 \, \mathrm{e}^{-W/kT} [\mathrm{A\,cm}^{-2}], \qquad (4.9)$$

where m_e is the rest mass of an electron and W the work function. It is found in practice that this equation gives a useful approximation but an overestimation of i_d for semiconductors. For usual values of W the dark current decreases strongly with decreasing temperature.

To eliminate the dc signals of background radiation and dark current the detection system operates in the alternating mode. However their frequency dependent fluctuations as shot noise will still mix with the signal.

Examples

Let us consider at room temperature the detection of $0.8\,\mu$m radiation with a GaAs–Cs$_2$O photocathode of $1\,\mathrm{cm}^2$ having $E_\mathrm{g} = 1.4\,\mathrm{eV}$, a negative affinity $\chi = -0.55\,\mathrm{eV}$ and a quantum efficiency of 15% for $0.8\,\mu$m radiation. The anode–cathode distance is $1.5\,$cm and the maximum applied voltage $500\,$V.

Using (4.9) with $W = E_\mathrm{g} + \chi = 0.85\,\mathrm{eV}$ we obtain $i_\mathrm{d} = 5.3 \times 10^{-8}$ $[\mathrm{A\,cm}^{-2}]$ and with the aid of (2.24) we find $\mathrm{NEP_{DL}} = 1.36 \times 10^{-12}$ $[\mathrm{W\,cm}^{-1}\,\mathrm{Hz}^{-1/2}]$.

Before calculating the Johnson noise we note that although this noise increases with the resistance its NEP, as described by (2.21), decreases with the resistance. Thus, under the condition of taking full profit of the detection speed we like to maximize R. The maximum value is set by the maximum frequency response or cutoff frequency given by (4.7). We calculate $f_\mathrm{c} = 2.2 \times 10^8$ Hz. The RC-time constant of the current circuit that satisfy this frequency condition is $RC < 1/2\pi f_\mathrm{c}$. The capacitance of the diode is given by $C = \varepsilon_0 A/d = 5.9 \times 10^{-14}$ F so that the maximum value of R is $R_\mathrm{m} = 1/2\pi C f_\mathrm{c} = 12\,\mathrm{k}\Omega$. Using (2.21) we now find $\mathrm{NEP_{AL}} = 1.2 \times 10^{-11}$ $[\mathrm{W\,cm}^{-1}\,\mathrm{Hz}^{-1/2}]$ which is larger than the $\mathrm{NEP_{DL}}$. Applying (2.2) we find $D^* = 8.3 \times 10^{10}$ $[\mathrm{W}^{-1}\,\mathrm{cm}\,\mathrm{Hz}^{1/2}]$. In practice $\mathrm{NEP_{AL}}$ will be even larger because of the larger effective temperature by applying amplification and $\mathrm{NEP_{DL}}$ will be smaller because (4.9) is found to give an overestimation. The background noise for radiation with $\lambda < 0.8\,\mu$m is negligible compared with the Johnson noise.

In the case the more common photocathodes like (Cs)Na$_2$KSB, type S-20, or Cs$_3$Sb, type S-17, is used with a work function above $1.5\,$eV, a much smaller dark current will be found. For instance a S-20 cathode with $W = 1.55\,\mathrm{eV}$ gives at room temperature $i_\mathrm{d} = 8.7 \times 10^{-20}$ $[\mathrm{A\,cm}^{-2}]$. Assuming also a quantum efficiency of 15% the corresponding $\mathrm{NEP_{DL}} = 1.74 \times 10^{-18}$ $[\mathrm{W\,cm}^{-1}\,\mathrm{Hz}^{-1/2}]$.

In conclusion we mention that the vacuum photodiodes are amplifier limited. With decreasing work function of the photon emitter the radiation sensitivity moves to the infrared wavelength region, although the dark current increases.

4.2 Photomultiplier

The noise equivalent power of the vacuum diode, mainly determined by the amplifier noise, can be considerably reduced if the photocurrent is amplified prior to the load resistor. This is accomplished in a photomultiplier by a process of secondary electron emission. In this device the initial photocurrent is accelerated by an electrostatic field between a series of electrodes, called dynodes, in which secondary electrons are a multiple of the incident electrons colliding with the dynodes. The dynodes are kept at progressively higher potential with respect to the photocathode, with typical potential intervals between dynodes of about 100 V. The last electrode, the anode, collects in this way the amplified current. Thus the initial photoelectrons gain energy from this acceleration so that the impact of each electron with a dynode releases multiple secondary electrons. The repeated process results in a gain of more than one million at the anode and produces a measurable current from a very small incident photon flux. Under optimum conditions the photomultiplier can even count single photons.

The whole amplification process occurs within a vacuum envelope to avoid interactions with gases. Usually the electron beam is focused by shaping the electrostatic field. The photocathode materials which determine the wavelength response are the same as applied for the vacuum photodiodes. Because of the large number of emitting surfaces the dark current produced by thermionic emission may limit the performance.

A schematic diagram of such a structure is shown in Fig. 4.6. Most commercially available systems have dynodes with emission surfaces of magnesium oxide or beryllium oxide which both have low thermionic emission and provide moderate gains of 3–5 per stage for acceleration voltages in the range of 100 V. Surfaces of Cs_3Sb or GaP have much higher secondary electron emission. Surfaces with negative electron affinity like GaAs–Cs_3O deliver with higher accelerating voltages even gains of 20–50 per stage, although the dark current may increase also remarkably. Construction details are found in reference [17].

The current gain of the photomultiplier is G and if the gain per stage is constant this gain can be described by $G = \delta^N$ where δ is the average gain

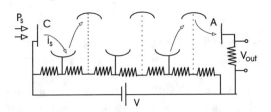

Fig. 4.6. Principle of a photomultiplier

per stage and N the number of stages. For instance, if $\delta = 5$ and $N = 10$ an overall gain of 10^7 is reached. Most commercially available photomultipliers have current gains in the range of 10^7–10^8. Using (2.3) the output signal current becomes

$$i_s = \frac{e\eta P_s}{h\nu}G\,, \tag{4.10}$$

where the quantum efficiency η is typically 10–25% of the incident photons over a narrow spectral range. For low signal power the linearity of these devices is excellent. At high signal powers, depending on specifications, saturation effects due to space charges occur. The time interval between photoemission and output signal, mostly a few nanoseconds, depends on the dynode constructions. The transit time spread of a photomultiplier is usually in the range of a few nanoseconds whereas high performance tubes exhibit very stable gains within a few percent.

Not only the signal current but also its noise given by (2.4) is amplified. The mean square of the amplified input signal noise current becomes

$$\overline{i_n^2} = 2ei_sG^2B\,. \tag{4.11}$$

Beside this amplified input noise the multiplication of the signal current by each stage introduces additional noise which on its turn is further amplified by the subsequent stages. This is because the electron emission events at a dynode are randomly distributed in time, although the average current amplification is described by the factor δ. This amplification noise can be calculated approximately by assuming Poisson statistics for the emission events. In the following we take for simplicity the average gain δ the same for each stage.

The amplified signal current after the first stage is δi_s and introduces a noise current equal to $\overline{i_{n1}^2} = 2e\delta i_s B$ which after subsequent multiplication by the following $N-1$ stages gives an output noise

$$\overline{i_{n1,\text{out}}^2} = 2e\delta i_s B\delta^{2N-2}\,. \tag{4.12}$$

For the second dynode we obtain similarly a signal current $\delta^2 i_s$, a noise current $\overline{i_{n2}^2} = 2e\delta^2 i_s B$, and an output noise

$$\overline{i_{n2,\text{out}}^2} = 2e\delta^2 i_s B\delta^{2N-4}\,. \tag{4.13}$$

Continuing for all dynodes and summing up with including the amplified input signal noise given by (4.11) we get

$$\overline{i_{n,s}^2} = 2ei_s B\left(\delta^{2N} + \sum_{n=1}^{N}\delta^n\delta^{2N-2n}\right) \tag{4.14}$$

or

$$\overline{i_{n,s}^2} = 2ei_s\delta^{2N}B\left(1 + \sum_{n=1}^{N}\delta^{-n}\right). \tag{4.15}$$

The term within the brackets is equal to $\Gamma = (\delta - \delta^{-N})/(\delta - 1)$ which is about $(\delta)/(\delta - 1)$ for $\delta^N \gg 1$. A typical example being $\Gamma = 1.33$ for $\delta = 4$. Substituting (2.3) the amplified signal noise can finally be written as

$$\overline{i_{n,s}^2} = \frac{2e^2\eta P_s}{h\nu}\Gamma G^2 B. \tag{4.16}$$

Thus the input signal noise is at the output amplified by a factor ΓG^2 and the number of dynodes has practically no effect on the noise production. The amplification noise of the incident background radiation and dark current due to thermionic emission of the photocathode can be derived similarly.

Using (2.18) we obtain for the output background noise

$$\overline{i_{n,b}^2} = 2e^2\eta n\Gamma G^2 B \tag{4.17}$$

and by using (2.22) for the dark current

$$\overline{i_{n,d}^2} = 2ei_d\Gamma G^2 B. \tag{4.18}$$

The Johnson noise, produced by the load resistor of the photomultiplier, is not amplified and given by

$$\overline{i_{n,j}^2} = \frac{4kTB}{R}. \tag{4.19}$$

The signal-to-noise ratio, derived from the ratio of the square of the signal current given by (4.10) and the sum of the noise contributions given by the (4.16–4.19) becomes

$$\frac{S}{N} = \frac{P_s^2}{\frac{2h\nu_s\Gamma B}{\eta}\left[P_s + h\nu_s n + \frac{h\nu_s i_d}{e\eta} + \frac{2h\nu_s kT}{\Gamma G^2 e^2 \eta R}\right]}. \tag{4.20}$$

It is seen that the Johnson noise, including the amplifier noise when the effective temperature T_{eff} is applied, is reduced by the factor G^2. Because of the large values of G its contribution can be neglected in practice, even when relatively small values for R are used to obtain high frequency response. By eliminating the amplifier noise the photomultiplier will be limited by the dark current noise which includes also thermionic emission of the dynodes, field emission and positive ion generation due to residual gases. Therefore, high quality photomultipliers are high vacuum tubes containing dynode surface materials with low thermionic emission.

Depending on the position of the dynode a microcurrent pulse produced by a thermionic emission event at that dynode is less amplified than a micropulse starting from an emission event at the photocathode. This difference in current pulse offers the possibility to eliminate the smaller unwanted dark current pulses by operating in the so called photon-counting mode (see Chap. 9). In this operating mode a fast electronic circuit at the photomultiplier output identifies all output pulses above an adjustable threshold value, so that microcurrent pulses due to photon emission will pass and the smaller micropulses due to the thermal emission of dynodes or otherwise associated with photon emission will be stopped. This mode can of course only operate at low signal power for which the individual photon pulses do not overlap. By means of this discrimination the signal-to-noise ratio is further increased and may now be limited by the background or in the case of sufficient background shielding by the signal itself. The NEP for photon counting with signal limitation is then obtained from (4.20) and given by

$$\mathrm{NEP_{SL}} = \frac{2h\nu\varGamma B}{\eta} . \tag{4.21}$$

Although in practice the individual micro-output pulses of a low power beam can be clearly observed the calculated $\mathrm{NEP_{SL}}$ of the photomultiplier for this technique does not make sense. After all the $\mathrm{NEP_{SL}}$ defines that on the average one photon is detected in the observation time. The requirement of clearly maintaining the separation of the micropulses during the amplification process is not consistent with this definition of the $\mathrm{NEP_{SL}}$. Therefore the observation time interval must be increased. This can be done by counting or averaging the pulses over a much longer period than the time constant of the photomultiplier so that the average output power is obtained. The bandwidth corresponding with this observation period can then be applied in (4.21) to give meaning to the $\mathrm{NEP_{SL}}$. In this way the calculated $\mathrm{NEP_{SL}}$ is, of course, lower than the low power beam to be investigated.

Example

Let us consider a low power beam of 10^{-12} [W] of 0.8 μm wavelength. The beam is measured by photon counting with a photomultiplier having a bandwidth of $B = 1\,\mathrm{GHz}$, $\varGamma = 1.3$, and $\eta = 0.15$. The signal noise is dominating. Further, the output of the photomultiplier is measured with an adjustable narrow band storage circuit. What is the highest frequency of the beam variations that can be measured with a signal-to-noise ratio of 100?

The photon energy of the radiation is $h\nu = 2.47 \times 10^{-19}\,\mathrm{J}$. The microcurrent pulses at the output of the photomultiplier have a duration of $\tau = 1/2B = 0.5\,\mathrm{ns}$. The average separation time between two photons is $h\nu/P_\mathrm{s} = 2.5 \times 10^{-7}\,\mathrm{s}$ which is 500 times longer than the pulse duration. For $S/N = 100$ we have the condition that $P_\mathrm{s} = 100 \times 2h\nu\varGamma B/\eta = 4.3 \times 10^{-16} B$. Substituting $P_\mathrm{s} = 10^{-12}$ [W] we find the highest frequency $B = 2.3 \times 10^3\,\mathrm{Hz}$.

As we discussed the secondary-emission amplification makes it possible to eliminate the amplification noise. High quality devices, operating at low temperature with very low dark currents, may even approach the signal limited detection i.e., performance limited only by the statistics of the photoelectrons. Therefore for many applications the photomultiplier is the most practical or sensitive detector available. A very sensitive and versatile device in the ultraviolet, visible, and near-infrared regions of the electromagnetic spectrum. The extremely fast time response and rise times as short as a fraction of a nanosecond provide a measurement capability in a large number of applications that is by far superior to those of other detectors.

5

Semiconductor Photodetectors

Semiconductor devices have many attractive advantages and are often preferable to vacuum photodetectors. They are not only simpler, cheaper, and applicable in low-voltage integrated circuits but also their wavelength sensitivity into the infrared region have extended the detection spectrum range considerably. In principle, the semiconductor devices can be divided into photoconductors, photodiodes, and avalanche photodiodes with internal current gain similar to the photomultiplier.

The photoconductive device is composed of a uniform semiconductor material in which electron–hole pairs are created by the absorption of radiation. The observed change of conductivity produced by the creation of free carriers is a measure of the incident radiation. For the photodiode it is the change of its p–n junction in which photoinduced carriers are created that modify the current–voltage characteristic so that the incident radiation can be measured. At high reverse-biased voltages the induced carriers gain sufficient energy to produce new electron–hole pairs through ionization. In this way an avalanche signal current can be obtained.

5.1 Photoconductors

There are three different types of photoconductors which are based on intrinsic absorption, extrinsic absorption, and free carrier absorption, respectively. The principle of intrinsic photoconductivity is schematically illustrated in Fig. 5.1a, where the incident photon excites an electron from the valence band into the conduction band and so producing an electron–hole pair. In general the electron mobility is dominant so that the increased conductivity comes mainly from the increased, photoinduced, electron density.

The free carriers will eventually recombine, but until this occurs the conductivity is increased. The recombination, characterized by the life time of the free carriers, determines the maximum detection frequency. It is evident that the energy of a photon must be greater than the intrinsic energy gap so

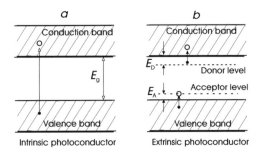

Fig. 5.1. Principle of photon absorption in a semiconductor

that the long-wavelength-limit to the photoconductivity λ_c is inversely proportional to the energy gap E_{g} or

$$\lambda_c E_{\mathrm{g}} = 1.24\,[\mathrm{eV}\,\mu\mathrm{m}]\,. \tag{5.1}$$

Roughly speaking the intrinsic photoconductors covers the spectral region between 0.53 μm radiation with CdS and about 25 μm with HgCdTe.

$\mathrm{Hg}_{1-x}\mathrm{Cd}_x\mathrm{Te}$ is mostly used for infrared imaging with arrays near $\lambda = 10\,\mu\mathrm{m}$. Its composition can be adjusted to get for instance near room temperature the optimized band gap for a photon at $\lambda \sim 10\,\mu\mathrm{m}$. Operating at $77\,\mathrm{K}$ this semiconductor is with its short life time very attractive in the developing technology of infrared imaging.

The semiconductors with small band gaps, responding to long wavelengths, must be cryogenically cooled otherwise the present thermally created free carriers swamp any small effect of photoinduced conductivity.

For larger wavelength, say in the region of 10–100 μm, extrinsic photoconductors with impurity states located in the band gap are applied. The physics of photon absorption in extrinsic semiconductors is shown in Fig. 5.1b. Semiconductors for which electrons can be excited from the valence band to their acceptor levels have p-type conductivity, whereas impurity-doped semiconductors for which photoexcited electrons are transferred from their donor states to the conduction band have n-type photoconductivity. As indicated in Fig. 5.1b a photon with energy $h\nu > E_{\mathrm{A}}$ excites an electron to the impurity level, leaving a hole in the valence band and thereby creating a change of the conductivity. Similarly for n-type photoconductivity a photon with energy $h\nu > E_{\mathrm{D}}$ transfers an electron from the impurity (donor) level into the conduction band. Usually the acceptor and donor levels are close to the valence and conduction band, respectively, so that the long-wavelength-limit for photon detection is considerably increased. For instance germanium has an intrinsic energy gap of about $0.68\,\mathrm{eV}$ ($\lambda_c = 1.8\,\mu\mathrm{m}$) but has with gold doping (p-type) an acceptor level that is only $0.15\,\mathrm{eV}$ ($\lambda_c = 8.3\,\mu\mathrm{m}$) above the valence band. To void thermal excitation the operating temperature must be below $60\,\mathrm{K}$. Germanium with copper doping has an acceptor level of only $0.041\,\mathrm{eV}$

Table 5.1. Band gap energy, long-wavelength-limit, and life time of excited state for various intrinsic and extrinsic semiconductors

semiconductor	T (K)	E_g (eV)	λ_c (µm)	τ (life time)
intrinsic				
Si	295	1.12	1.1	50 ps
Ge	295	0.68	1.8	10 ns
PbS	295	0.46	2.7	0.1–1 ms
PbS	195	0.4	3.1	1–10 ms
PbS	77	0.32	3.8	1–10 ms
PbSe	295	0.31	4	1–10 µs
PbSe	195	0.29	4.3	10–100 µs
PbSe	77	0.24	5.2	10–100 µs
PbTe	295	0.41	3	1–10 µs
PbTe	77	0.27	4.5	10–100 µs
CdTe	295	1.55	0.8	
CdSe	295	1.8	0.67	10 ms
CdS	295	2.4	0.53	50 ms
InSb	77	0.22	5.5	1–10 µs
$Hg_{0.8}Cd_{0.2}Te$	77	0.1	10–25	< 1 µs
extrinsic				
Ge:Au	77	0.15	8.3	30 ns
Ge:Cu	15	0.041	30	0.5 ns
Ge:Cd	20	0.06	21	10 ns
Ge:Zn	4	0.032	40	10 ns
Si:Ga	4	0.073	17	1 µs
Si:As	20	0.056	22	0.1 µs

($\lambda_c = 30$ µm) above the valence band and an operating temperature of 15 K. The cutoff wavelength extends to about 40 µm for germanium with zinc doping operating at 4 K. Examples of intrinsic and extrinsic semiconductors are given in Table 5.1.

It has been found experimentally that the long-wavelength-limit of extrinsic photoconductors is at about 100 µm. Nevertheless, longer wavelength sensitivity has been obtained with pure semiconductors such as Ge and InSb at low temperature by changing the conductivity with selective electron heating [18]. Normally at room temperature the coupling between lattice and electrons is so strong that if a static or alternating electric field is applied, that interacts mainly with the free electrons, the energy of the electrons will not be significantly greater than in the thermal equilibrium with the lattice. However, in pure high mobility semiconductors at cryogenic temperatures the coupling between electrons and lattice becomes so weak that even for quite small electric fields the steady-state energy of the free carriers is appreciable greater than the thermal equilibrium value with the lattice. Since the carrier mobility depends on its mean energy there is a relation between the mean carrier energy and the conductivity. By drawing energy from the radiation field as well as from a static field the changes of the radiation field can be observed through changes in the static conductivity. This principle of hot carrier effect has been successfully used over a wavelength range from $50\,\mu\text{m}$ to $10\,\text{mm}$. It is clear that the physics involved is not photon but energy dependent so that in fact we are dealing with thermal detectors having high sensitivity and small time constants, less than $1\,\mu\text{s}$.

5.1.1 Analysis of the Detection Process

When a voltage V is applied across a sample of semiconductor as depicted in Fig. 5.2 the generated current i_d, usually called bias or dark current, is equal to $i_d = V/R$ where R is the resistance of the material. Next to this bias current we apply a uniform illumination with power P_s of the sample as shown in the figure and we want to determine the change of the current which is due to this illumination. The applied voltage V is between transverse contacts on opposite faces as shown in the figure where w, l, and t are the width, length, and thickness of the sample. In practice the contacts are made of evaporated metal film. The current is then uniform through the cross-section wt. In the following we assume that the radiation power is uniformly transmitted through the cross-section wd.

The incident power with quantum efficiency η produces $\eta P_s/h\nu$ electron–hole pairs per second provided the photon energy is sufficient to ionize. If the recombination time is τ there will be a steady state of N electron–hole pairs given by

$$N = \frac{\eta P_s \tau}{h\nu} \, . \tag{5.2}$$

Per unit volume of the sample we get

$$n = \frac{N}{wtd} \, . \tag{5.3}$$

The photoconductance G_{ph} of the sample is given by

$$G_{ph} = e\left(\mu_e n_e + \mu_p n_p\right) \frac{wt}{d} \, , \tag{5.4}$$

where μ_e, n_e and μ_p, n_p are the mobility and density of, respectively, the electrons and the holes. Usually the mobility of one type of carriers dominates.

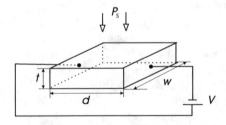

Fig. 5.2. Photon current circuit in a semiconductor

Particularly for intrinsic and n-type doped material the electron mobility is much larger than the hole mobility. For those materials we may replace (5.4) by

$$G_{\mathrm{ph}} = \frac{ne\mu wt}{d} .$$ (5.5)

Substituting (5.3) the photoconductance becomes

$$G_{\mathrm{ph}} = \frac{e\mu N}{d^2} ,$$ (5.6)

so that the photon current i_{s} will be

$$i_{\mathrm{s}} = G_{\mathrm{ph}} V = \frac{e\mu NV}{d^2} .$$ (5.7)

Substituting (5.2) and $\frac{\mu V}{d} = \mu E = v_{\mathrm{d}}$ into (5.7) we find

$$i_{\mathrm{s}} = \frac{e\eta P_{\mathrm{s}}}{h\nu} \frac{\mu \tau_{\mathrm{l}} V}{d^2} = \frac{e\eta P_{\mathrm{s}}}{h\nu} \left(\frac{\tau_{\mathrm{l}}}{\tau_{\mathrm{d}}} \right) = \frac{e\eta g P_{\mathrm{s}}}{h\nu} ,$$ (5.8)

where V is the voltage over the sample, $\tau_{\mathrm{d}} = d/v_{\mathrm{d}}$ the drift time of the carrier through the sample and $g = \tau_{\mathrm{l}}/\tau_{\mathrm{d}}$.

To measure this signal current i_s a biasing circuit with a load resistance R_{L} is used as shown in Fig. 5.3. It consists of the photoconductive detector element with resistance R which is in series with R_{L} connected to a constant voltage supply. The bias current through the circuit is $i_0 = V_0/R + R_{\mathrm{L}}$. The detector element is sufficiently thick to absorb the radiation. Intrinsic semiconductor elements are often thin films obtained by evaporation or chemically deposited on nonconductive material like glass or ceramics.

The incident radiation power P_{s} produces a current source generating i_s. The return current flows through the resistors R_{L} and R. In other words the detector can be considered as a current generator across which there are the detector and load resistances. The current parts through R_{L} and R are, respectively, i'_{s} and i''_{s}. The signal current i'_{s} through R_{L} is

$$i'_{\mathrm{s}} = \frac{R}{R + R_{\mathrm{L}}} i_{\mathrm{s}}$$ (5.9)

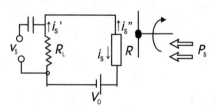

Fig. 5.3. Principle of photoconductive detector. The generated electron–hole pairs by the chopped incident radiation change the conductivity of the detector. The resulting change of the circuit current is observed as a voltage drop V_{s} over the load resistance R_{L}

and the voltage increase over R_{L} is

$$V_{\mathrm{s}} = \frac{R R_{\mathrm{L}}}{R + R_{\mathrm{L}}} i_{\mathrm{s}} \, . \tag{5.10}$$

It is seen that the maximum signal response[1] is obtained for $R_{\mathrm{L}} \gg R$ or constant bias current. The signal current given by (5.8) increases linearly with V because v_{d} is proportional to V. The question is what is the optimum signal voltage over the detector element. To answer this question we have to consider also the noise which is derived by (1.44). According to this equation the g–r noise is proportional to the current and for weak signals, $i_{\mathrm{s}} \ll i_0$, this will be proportional to the bias current which is again proportional to V. The value of $\overline{i_{\mathrm{n}}^2}$ of the g–r noise as given by (1.44) is proportional to V^2 (because both i_0 and g are proportional to V) whereas i_{s}^2 is also proportional to V^2 so that S/N accordingly will be independent on V. However, it is observed that for the usual semiconductor elements the signal current increases less than linearly or the mean square root of the noise current increases more than linearly by increasing voltage. Further increase in voltage beyond this optimum value results in a decreased S/N value and if the voltage is increased much more the charge carriers will be accelerated sufficiently to produce additional carriers in collisions, leading to a breakdown avalanche process. It can also happen that the maximum voltage is set by the power dissipation in the detector element due to ohmic heating which influences the carrier life time and its mobility. The maximum power dissipation is usually considered about $0.1 \, \mathrm{W \, cm^{-2}}$ of element area.

The responsivity given by the ratio of the signal voltage and the input radiation power is obtained by substituting (5.8) into (5.10).

$$r = \frac{V_{\mathrm{s}}}{P_{\mathrm{s}}} = \frac{e \eta g R_v}{h \nu} \, , \tag{5.11}$$

where $R_v = R R_{\mathrm{L}}/(R + R_{\mathrm{L}})$. Since the responsivity depends on the photon energy it reaches a maximum for photon energies close to band gap energy or in the case of extrinsic material close to the minimum ionization energy. At smaller wavelength only a part of the photon energy is used for the transition and therefore the response decreases with wavelength.

Considering the detector noise we note that the Johnson noise produced by R_v and the g–r noise produced by the signal, background and bias currents will be divided over R and R_{L} similarly as the signal current. This means that for calculating the signal-to-noise ratio only the value of R_v is relevant and not the individual values of R and R_{L}. The spectral power density of the signal noise current is derived in Sect. 1.5 and is given by (1.44). The background noise current is given by (2.18). The conductivity of the bias (often called dark) current comes from the thermally excited carriers

---

[1] It is sometimes erroneously stated to bias the detector with a load resistance equal to the resistance of the detector for obtaining maximum signal response.

which also produce g–r noise. The average life time of those carriers may be different from the optically excited carriers. However, we assume that the average life times are equal so that also the g-factors are equal. Including also the Johnson noise the signal-to-noise ratio in terms of current can be written as

$$\frac{S}{N} = \frac{\left(\dfrac{ge\eta}{h\nu_s}\right)^2 P_s^2}{4ge\left(\dfrac{ge\eta P_s}{h\nu_s}\right)B + 4\eta\,(ge)^2\,Bn + 4gei_dB + \dfrac{4kT_{eff}B}{R_v}}, \qquad (5.12)$$

where the dark current is $i_d = i_0$. For optimum performance $R_L \gg R$ so that we replace R_v by R and obtain

$$\frac{S}{N} = \frac{P_s^2}{\dfrac{4h\nu_s B}{\eta}\left[P_s + h\nu_s n + \dfrac{h\nu_s i_d}{ge\eta} + \dfrac{h\nu_s kT_{eff}}{g^2 e^2 \eta R}\right]}. \qquad (5.13)$$

The signal limited NEP_{SL}, which is the ultimate in detector sensitivity, is by far much smaller than the limitations by the other noise sources. Thus signal limitation is not feasible with photoconductors. Then, noise due to the fluctuations of the background radiation may set a fundamental limit to the detectivity. Whether this can be reached depends on the noise of the dark current and amplifier. Let us consider the ratio of the dark current noise power to the thermal or Johnson noise power. Looking at (5.13) this ratio is

$$N_s = \frac{V e g}{kT_{eff}}, \qquad (5.14)$$

where we have substituted the detector resistance $R = V/i_d$. If we now substitute $g = \tau\mu V/d^2$ in the last equation we find

$$N_s = \frac{e\tau\mu}{kT_{eff}}\left(\frac{V}{d}\right)^2. \qquad (5.15)$$

Photoconductors operating with optimum bias voltage have often field strengths V/d in the order of $100\,\mathrm{V\,cm^{-1}}$. With this value it turns out that in general N_s is much larger than unity so that the dark current noise is dominating. From (5.13) we then find the dark current limited NEP_{DL} as

$$NEP_{DL} = \frac{2h\nu}{\eta}\sqrt{\frac{i_d B}{ge}} = \frac{2h\nu}{\eta}\sqrt{\frac{Bd^2}{\tau\mu eR}}. \qquad (5.16)$$

Considering a square sensitive detector element with area equal to d^2 we obtain by using (2.2) for the specific dark current limited detectivity $D_{DL}^* = \frac{\eta}{2h\nu}\sqrt{\tau\mu eR}$. The resistance R can be expressed (see Fig. 5.2) as $R = (d/wt)\rho$, where ρ is the specific resistance of the semiconductor. The minimum value

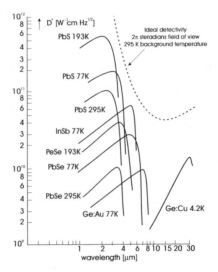

Fig. 5.4. The specific dark current limited detectivity as a function of wavelength for various photoconductors is plotted and compared with the ideal detectivity

of t is determined by the photon absorption coefficient. By taking $d = w$ and $t = 1/\alpha$, where α is the absorption coefficient we get for the specific detectivity

$$D^*_{\mathrm{DL}} = \frac{\eta}{2h\nu}\sqrt{\tau\mu e\rho\alpha}\,. \tag{5.17}$$

It is seen that the specific detectivity is only determined by material properties and is *independent* on the bias current and the dimensions of the photoconductor. In Fig. 5.4 the variation of D^*_{DL} with wavelength is plotted for various semiconductors.

Examples

1. A film type intrinsic lead selenide detector with an absorption coefficient $\alpha = 2 \times 10^4\,\mathrm{cm}^{-1}$ has an efficiency of 65%. The film with a square sensitive area and a thickness of a few microns has a resistance of $5 \times 10^4\,\Omega$. It is used at room temperature to detect $3\,\mu\mathrm{m}$ radiation. At room temperature the mobility $\mu = 10\,\mathrm{cm}^2\,\mathrm{V}^{-1}\,\mathrm{s}^{-1}$ and $\tau = 10^{-6}\,\mathrm{s}$. The optimum bias voltage is $200\,\mathrm{V\,cm}^{-1}$.

   Applying (5.15) we find $N_\mathrm{s} = 15$ so that the system is dark current limited. Using (5.16) we get $\mathrm{NEP}_{\mathrm{DL}} = 7.27 \times 10^{-10}(Bd^2)^{1/2}\,\mathrm{W}$. The specific detectivity according to (2.2) is $D^* = 1.37 \times 10^9\,\mathrm{W}^{-1}\,\mathrm{cm\,Hz}^{1/2}$, which is much smaller than $D^*_i(3\,\mu\mathrm{m})$ of about 10^{12}. Thus this system is dark current limited [19].

2. An HgCdTe detector with a square sensitive area of $2.5 \times 2.5\,\mathrm{mm}^2$ operating at liquid nitrogen temperature is used to detect $10\,\mu\mathrm{m}$ radiation [20]. The field of view of the detector element is $60°$. The material parameters are: $\tau = 1.2 \times 10^{-6}$, $\mu = 10^4\,\mathrm{cm}^2\,\mathrm{V}^{-1}\,\mathrm{s}^{-1}$, $V/d = 100\,\mathrm{V\,cm}^{-1}$, $R = 38\,\Omega$, $\eta = 80\%$.

With (5.15) we find for $T = 77\,\mathrm{K}$ that $N_\mathrm{s} = 1.6 \times 10^4$ so that the Johnson noise is negligible. (If an effective temperature T_eff much higher than $77\,\mathrm{K}$ is substituted in order to include also the amplifier noise, this noise is still negligible.) Using (5.16) we obtain $\mathrm{NEP_{DL}} = 2.7 \times 10^{-10}(Bd^2)^{1/2}\,\mathrm{W}$ and with (2.2) $D^* = 3.7 \times 10^9\,\mathrm{W^{-1}\,cm\,Hz^{1/2}]}$, which is more than a factor 10 smaller than D_i^*. Thus the system is dark current limited.

5.1.2 Frequency Response

The minimum capacitance of the sensitive element is determined by the photon absorption coefficient. If the reflectivity is negligible due to an anti-reflection layer the absorption coefficient is $\eta = (1 - \mathrm{e}^{-\alpha t})$ where t is the thickness. By taking $t = 1/\alpha$ the efficiency is already 63%. Considering square sensitive elements the capacitance of a device becomes $C = \varepsilon_\mathrm{r}\varepsilon_0/\alpha$. The absorption coefficient of most intrinsic materials varies from 10^3 to $10^4\,\mathrm{cm^{-1}}$ and ε_r is in the range of 10–20 so that the minimum capacitance of an element is roughly 10^{-15}–$10^{-16}\,\mathrm{F}$ which is in practice negligible compared with the stray capacitance and the capacitance of the leads in a circuit. It turns out that the RC time constant is very small and usually much smaller than τ_l so that the frequency response is set by this recombination time.

For extrinsic material the absorption coefficient is much less, in the range of 1–$10\,\mathrm{cm^{-1}}$. Nevertheless the resulting RC time is mostly still smaller than τ_l and the frequency response is still limited by the recombination time. However, devices may have a frequency limitation by the bandwidth of the preamplifier (operational amplifier) or stray capacitance of the leads into the dewar containing the detector.

5.2 Photodiodes

The semiconductor p–n junctions, called diodes, are widely used as photodetectors. The junction is obtained when a piece of p-type material is in contact with a piece of n-type material. Near the junction a depletion layer is formed in which the impurity atoms are fully ionized. In fact a double layer is formed with equal amounts of positive (near the boundary of the n-type) and negative charges (near the boundary of the p-type). Within the double layer a strong electric field exist. If photogenerated charge carriers are formed near the junction the double layer will separate very fast the negative and positive charge carriers. In this way a photocurrent is generated. In a sense the depletion layer behaves like a vacuum diode.

Because effective photocurrent generation can only take place for carriers created near or in the depletion region, as we shall see, high radiation absorption is required. This can be obtained with intrinsic transition. Extrinsic transitions with their much smaller absorption coefficients are in general not suitable. For this reason the detectable wavelengths of photodiodes are smaller than those of photoconductors. Widely used photodiodes with their dependences of detectivity on wavelengths are shown in Fig. 5.5. The detectivities of the photodiodes are dark current limited.

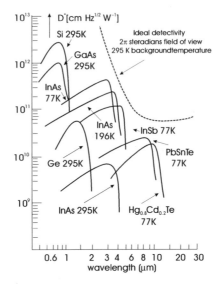

Fig. 5.5. The specific dark current limited detectivity as a function of wavelength for various reverse-biased diodes is plotted and compared with the ideal detectivity

Longer wavelength detection is feasible due to relatively modern development of mixed alloys like $Pb_{1-x}Sn_xTe$, $Pb_{1-x}Sn_xSe$, and $Hg_{1-x}Cd_xTe$ whose gap energy can be constructed by varying the ratio of two major components.

The relatively high shot noise of photodiodes, typically three orders of magnitude higher than that of the photomultiplier, limits their ability to detect low light levels.

Before discussing the physics of the created charge carriers and the subsequent photocurrent it is helpful to review the physics of the p–n junction.

5.2.1 P–N Junction

In Fig. 5.6a a p–n junction with electrical contacts is shown. We assume for simplicity an abrupt change of donor to acceptor doping at the contact surface. As discussed in Sect. 5.1 the Fermi level in the p-type is located close to the valence band and the Fermi level of the n-type is close to the bottom of the conduction band as shown in Fig. 5.6b. As a consequence the electrons in the vicinity of the junction diffuse from the n-type into the p-type material until the Fermi levels are equal.

The electrons will combine with the holes producing a space charge region of negative charges in the p-type and positive charges in the n-type material with a potential drop V_0. This is indicated in Fig. 5.6c. The double layer space charge region deprived of its carrier conductors exhibits high resistance. If we now apply an external voltage V_b between the contacts the potential drop appears at the junction as indicated in Fig. 5.6d where the voltage is reverse biased (a positive voltage V applied to the n-type side of the junction relative to the p-type side). By increasing the potential barrier across the junction

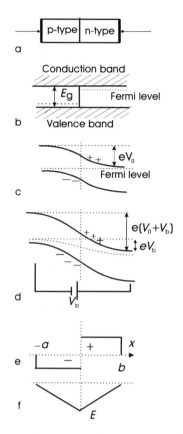

Fig. 5.6. (**a**) Junction between a p-type and an n-type semiconductor with electrical contacts. (**b**) Band gaps, gap energy E_{g}, and Fermi levels (*dotted lines*) of the p- and n-type semiconductor in the absence of contact between the two materials. (**c**) Change of band levels by contact due to space charges near the boundary. The Fermi levels are equal. (**d**) Change of band and Fermi levels due to space charges and an applied reverse biased voltage over the junction. (**e**) Double space charge layer near the junction. (**f**) Electric field distribution of the double space charge region

from eV_0 to $e(V_0 + V_{\mathrm{b}})$ the double layer will increase. Its thickness can be calculated by assuming that all donors in the region between $x = 0$ and $x = b$ and all acceptors in the region between $x = -a$ and $x = 0$ are ionized as shown in Fig. 5.6e. Using the Poisson equation we have

$$\frac{\mathrm{d}^2 V}{\mathrm{d}x^2} = -\frac{eN_{\mathrm{D}}}{\varepsilon} \quad \text{for} \quad 0 < x < b \tag{5.18}$$

and

$$\frac{\mathrm{d}^2 V}{\mathrm{d}x^2} = +\frac{eN_{\mathrm{A}}}{\varepsilon} \quad \text{for} \quad -a < x < 0, \tag{5.19}$$

where N_{D} and N_{A} the donor and acceptor densities, respectively, ε the dielectric constant and $-e$ the charge of an electron. Solving (5.18) and (5.19) we apply, as indicated in Fig. 5.6f, the boundary conditions $E = -\frac{dV}{dx} = 0$ for $x = -a$ and $x = b$ and the continuity of V for $x = 0$. We obtain

$$V = \frac{e}{2\varepsilon} N_A \left(x^2 + 2ax\right) \quad \text{for} \quad -a < x < 0, \tag{5.20}$$

$$V = \frac{-e}{2\varepsilon} N_D \left(x^2 - 2bx\right) \quad \text{for} \quad 0 < x < b. \tag{5.21}$$

Because of charge neutrality the total charges of each layer are opposite and equal so that

$$aN_A = bN_D. \tag{5.22}$$

The total voltage drop over the depletion layer $V_0 + V_b = V(b) - V(-a)$ becomes

$$V_0 + V_b = \frac{e}{2\varepsilon} \left(N_D b^2 + N_A a^2\right). \tag{5.23}$$

Solving for the width of the space charge region $w = a + b$ we obtain from the two last equations

$$w = \left[\frac{2\varepsilon}{e} \left(V_0 + V_b\right) \frac{N_A + N_D}{N_A N_D}\right]^{1/2}. \tag{5.24}$$

Thus the space charge region increases with the reverse-biased voltage V_b.
The diode capacitance C_d is given by

$$C_d = \frac{\varepsilon A}{w}, \tag{5.25}$$

where A is the contact area of the junction. It should be note that the capacitance decreases with increasing reverse-biased voltage as is seen from (5.24).

5.2.2 Current–Voltage Characteristic

Next we calculate the current–voltage characteristic for the p–n junction. In the absence of a bias voltage the equilibrium relation between the minority carrier concentration on one side of the space charge region and the majority carrier concentration on the other side is given by the Boltzmann law or

$$p_n = p_p \, e^{-eV_0/kT}, \tag{5.26}$$

where p_n is the hole density of the n-type material at the boundary of the space charge region and p_p the hole density of the p-type material. Similarly the relation between the free electron density n_p of the p-type material at the boundary of the space charge region and its density of the n-type material is

$$n_p = n_n \, e^{-eV_0/kT}. \tag{5.27}$$

Away from the boundaries the minority carriers will gradually change to their equilibrium bulk values. Although the minority carriers near the boundaries have gradients there is no net current because the carriers experience also an electrostatic force which drives them in the opposite direction of their diffusion current.

The equilibrium relations given by (5.26) and (5.27) are the result of the balance between the continuous thermal creation of electron–hole pairs and the recombination of the charge carriers. If g is the production rate of pairs per unit volume we have at equilibrium in the p-type material $g_p = r_p p_p n_p$, where r_p is the recombination coefficient. Similarly for the n-type material we have $g_n = r_n p_n n_n$. Substitution of (5.26) and (5.27) shows that the ratio g/r is independent of donor or acceptor concentration. Then g/r is also equal to n_i^2, where n_i is the density of an intrinsic carrier. Thus we have the relations

$$p_p n_p = p_n n_n = n_i^2 . \tag{5.28}$$

Next we consider the effect of a positive voltage V applied to the p-type side of the junction relative to the n-type side (forward biased). Because of the high resistance of the space charge region the voltage drop is across this depletion layer so that the bias voltage has practically no effect on the majority carriers p_p and n_n on both sides. At equilibrium the minority carriers, however, n_p^V and p_n^V at the boundaries of the space charge region now change according to the Boltzmann law as

$$p_n^V = p_p \, e^{-e(V_0-V)/kT} = p_n \, e^{eV/kT} , \tag{5.29}$$

$$n_p^V = n_n \, e^{-e(V_0-V)/kT} = n_p \, e^{eV/kT} , \tag{5.30}$$

where we have used (5.26) and (5.27). The voltage-induced minority carriers $\Delta p_n = p_n^V - p_n$ and $\Delta n_p = n_p^V - n_p$ will now diffuse away from the space charge region and reach the contacts. The distributions of the charge carriers for $V > 0$ are shown in Fig. 5.7.

The voltage-induced minority carriers generate a diffusion current. The equation describing their diffusion and recombination in a stationary process is for the holes

$$D_p \frac{d^2 p_n^V}{dx^2} = \frac{p_n^V - p_n}{\tau_p} , \tag{5.31}$$

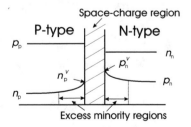

Fig. 5.7. Distribution of the minority charge carriers of a forward biased p–n junction

where τ_{p} is the recombination time and D_{p} the diffusion coefficient for holes. The solution is

$$p_{\mathrm{n}}^{V}(x) = \left[p_{\mathrm{n}}^{V}(0) - p_{\mathrm{n}}\right] e^{-x/L_{\mathrm{p}}} + p_{\mathrm{n}} . \tag{5.32}$$

Similarly the diffusion and recombination of the electrons in the p-type material is for stationary conditions described by

$$D_{\mathrm{n}}\frac{\mathrm{d}^2 n_{\mathrm{p}}^{V}}{\mathrm{d}x^2} = \frac{n_{\mathrm{p}}^{V} - n_{\mathrm{p}}}{\tau_{\mathrm{n}}} , \tag{5.33}$$

where τ_{n} is the recombination time and D_{n} the diffusion coefficient for electrons. The solution is

$$n_{\mathrm{p}}^{V}(x) = \left[n_{\mathrm{p}}^{V}(0) - n_{\mathrm{p}}\right] e^{+x/L_{\mathrm{n}}} + n_{\mathrm{p}} , \tag{5.34}$$

where x is negative in the p region. $L_{\mathrm{p}} = \sqrt{D_{\mathrm{p}}\tau_{\mathrm{p}}}$ and $L_{\mathrm{n}} = \sqrt{D_{\mathrm{n}}\tau_{\mathrm{n}}}$ are the diffusion lengths for the holes and electrons, respectively.

The total current through the junction is equal to the flow across the space-charge region which is given by the diffusion currents of the minority carriers at the boundaries of the space-charge region. The current densities are according to the diffusion equations proportional to the negative gradient of the charge densities. For the electron current density we get

$$i_{\mathrm{n}} = +eD_{\mathrm{n}}\left(\frac{\mathrm{d}n_{\mathrm{p}}^{V}}{\mathrm{d}x}\right)_{x=0} , \tag{5.35}$$

where the current is positive in the direction of x. For the holes we get similarly

$$i_{\mathrm{p}} = -eD_{\mathrm{p}}\left(\frac{\mathrm{d}p_{\mathrm{n}}^{V}}{\mathrm{d}x}\right)_{x=0} . \tag{5.36}$$

Substituting (5.32) and (5.34) into (5.35) and (5.36) and substituting in the result (5.29) and (5.30) we obtain

$$i_{\mathrm{n}} = \frac{eD_{\mathrm{n}}n_{\mathrm{p}}}{L_{\mathrm{n}}}\left(e^{eV/kT} - 1\right) \tag{5.37}$$

and

$$i_{\mathrm{p}} = \frac{eD_{\mathrm{p}}p_{\mathrm{n}}}{L_{\mathrm{p}}}\left(e^{eV/kT} - 1\right) . \tag{5.38}$$

The current–voltage characteristic of a diode is the sum of i_{n} and i_{p} multiplied by the contact area A or

$$i = i_{\mathrm{d}}\left(e^{eV/kT} - 1\right) , \tag{5.39}$$

where the saturation current i_d is given by

$$i_d = A \left(\frac{eD_n n_p}{L_n} + \frac{eD_p p_n}{L_p} \right). \tag{5.40}$$

This saturation or dark current is due to the thermal generation of carriers in the depletion layer.

It is of interest to calculate the diode noise. We derive for zero current the diode resistance from $\frac{1}{R_d} = \frac{di}{dV}$. With (5.39) we obtain

$$R_d = \frac{kT}{ei_d}. \tag{5.41}$$

The diode current in forward direction for $V = 0$ is equal to the reverse or saturation current i_d. The associated shot noises of the two currents are not correlated so that the total shot noise is $4ei_d B$. If we now eliminate i_d by means of (5.41) we get for the noise current $\frac{4kTB}{R_d}$ which is just the Johnson noise.

5.2.3 Photon Excitation

Charge carriers are continuously produced by thermal excitation and recombine in both regions of the junction. The thermally excited minority carriers will randomly vary their distributions also near the space-charge region and thereby effecting the diffusion currents i.e., the diffusion currents have a noise components added to their average values. The noise current generated by a thermally excited minority carrier depends on its trajectory during its life time before recombination. The closer to the space-charge region the larger its effect on the diffusion current. If they are created away from the space-charge where the diffusion is absent they are practically not driven and will recombine without generating noise current. The minority carriers produced in or at the boundary of the space-charge region will be driven very fast by the junction field and recombine on the other side. They deliver full charge transfer through the junction and thus producing noise.

In case of photon excitation the minority carriers behave the same as the thermally excited carriers. The additional photon excited minority carriers produced near the space-charge region decrease the diffusion currents so that they generate a reverse current. Then the photon efficiency depends on the position of the absorption. The further away from the space-charge the lower the efficiency. At the boundary or within the space-charge the efficiency is one because the transit time for the carriers through the junction due to the strong field is much smaller than the recombination time.

Analyzing this process quantitatively we refer to Fig. 5.8. Suppose that in the n-region of the junction electron–hole pairs are created by photon absorption within a thin sheet source at a distance x_0 from the space-charge. The distribution of the minority carriers can again be calculated by solving the

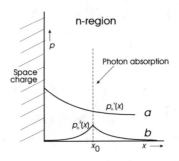

Fig. 5.8. Minority carrier distribution for photon excitation

diffusion equation. The linearity of this differential equation means that the solution is the sum of solutions describing the minority distributions caused by each of the individual sources and their boundary conditions. Thus the desired distribution can be written as the sum of the undisturbed distribution given by (5.32) and the disturbance by the sheet source. Therefore we are now calculating the part that is due to this sheet source. Looking at the diffusion equation it is seen that on the left side of the sheet source the solution is of the general form

$$p_n^s = c_1 \, e^{x/L_p} + c_2 \, e^{-x/L_p} \,. \tag{5.42}$$

Since at the boundary of the space-charge region the density of the minority carriers, given by (5.32), was already included in the undisturbed solution we now require that the solution of (5.42) is zero for $x = 0$ so that we find $c_1 = -c_2$. On the right of the sheet source, $x > x_0$, the solution has only a negative exponential function because $p_n^s(\infty) = 0$. So we have

$$p_n^s(x) = c_3 \, e^{-x/L_p} \,. \tag{5.43}$$

Applying the continuity at the thin sheet source we get $c_3 = c_1(e^{2x_0/L_p} - 1)$. Thus we find for the disturbance caused by the sheet source

$$p_n^s(x) = c_1 \left(e^{x/L_p} - e^{-x/L_p} \right) \quad \text{for} \quad 0 < x < x_0 \,, \tag{5.44}$$

and

$$p_n^s(x) = c_1 \left(e^{2x_0/L_p} - 1 \right) e^{-x/L_p} \quad \text{for} \quad x_0 < x < \infty \,, \tag{5.45}$$

as shown in Fig. 5.8.

The hole current caused by the photoionization source consists of two components: one to the left (i_L) and the other one to the right (i_R) of the source at $x = x_0$. Thus $i_R = -eD_p \frac{dp_n^s}{dx}$. It becomes for $x = x_0$

$$i_R(x_0) = \frac{eD_p}{L_p} c_1 \, e^{-x_0/L_p} \left(e^{2x_0/L_p} - 1 \right) \,. \tag{5.46}$$

Similarly $i_L = +eD_p \frac{dp_n^s}{dx}$ and becomes for $x = x_0$

$$i_L(x_0) = \frac{eD_p}{L_p} c_1 e^{-x_0/L_p} \left(e^{2x_0/L_p} + 1 \right) . \tag{5.47}$$

The total current from the source is then

$$i_T = i_R + i_L = \frac{2eD_p}{L_p} c_1 e^{x_0/L_p} . \tag{5.48}$$

The current through the junction due to the photoionization source is given by $i_J = -eD_p \frac{dp_n^s}{dx}$. Using (5.44) we obtain for $x = 0$

$$i_J = -\frac{2eD_p}{L_p} c_1 . \tag{5.49}$$

Since all quantities on the right hand side of the last equation are positive we thus find that the photocurrent is negative or *reverse*.

The quantum efficiency is defined as the ratio of the electron–hole pairs of which the minority carriers reach the junction and generate a photocurrent to the incident ionizing photons. For the considered thin sheet source we obtain from (5.49) and (5.48)

$$\eta = \frac{i_J}{i_T} = e^{-x_0/L_p} . \tag{5.50}$$

Thus for good efficiency the photon absorption should take place close to or at least within the diffusion length distance from the junction.

So far we have discussed the photon absorption at the n side of the diode. If the photons are absorbed at the p side the process is similar. The resulting electrons will also diffuse to the junction and drift across the high-field region of the junction, giving also rise to a reverse current in the external circuit. The quantum efficiency is then also given by (5.50) except that the diffusion length is replaced by L_n and x_0 is the distance in the p-side to the space-charge region.

Photons may also be absorbed in the depletion layer. The created holes and electrons will then be driven in opposite directions by the strong electric field of the junction and reach the p- and n-side. For these carriers the transit time is much shorter than the recombination time so that for the corresponding photons the efficiency is one.

A typical photodiode construction, shown in Fig. 5.9a consists of a thin p-layer, usually less than $1\,\mu m$, on the surface and an underlying n-layer. The diode is covered with an antireflection coating. The absorption through the top layer may be small so that the photons are mainly absorbed in the n-type material. To obtain the diode efficiency we may, because of the linearity of the diffusion equation, simply integrate (5.50) and obtain

$$\eta = \int_0^d \alpha e^{-\alpha x} e^{-x/L_p} \, dx , \tag{5.51}$$

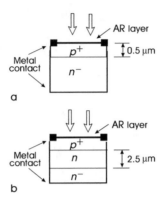

Fig. 5.9. (a) Photodiode construction having a thin p-type top layer covered with an antireflection coating. **(b)** Construction of a typical silicon PIN photodiode having a thin intrinsic layer between the p- and n-type layers

where due to the antireflection coating reflection losses are neglected. The absorption coefficient is given by α. The thickness of the n-layer is d. We obtain

$$\eta = \frac{\alpha L_{\mathrm{p}}}{\alpha L_{\mathrm{p}} + 1} \left(1 - \mathrm{e}^{-d(\alpha + 1/L_{\mathrm{p}})} \right) . \tag{5.52}$$

Good efficiency requires both d and L_{p} to be larger than $1/\alpha$. This condition for high absorption is usually not fulfilled for extrinsic transitions. Therefore photodiodes operate through intrinsic rather than through extrinsic absorption. Consequently the wavelength range for photodiodes is much smaller than for photoconductors and the overlap is on the short wavelength side.

Efficiency improvement is reached with p–i–n photodiodes, usually called PIN-diodes, in which a thin intrinsic layer is sandwiched between the p- and n-regions. The diode is illuminated through the p-layer which is again very thin so that its absorption is negligible. Since the intrinsic layer has high resistivity the potential drop over the junction is mainly over the i-layer and the created holes and electrons will be driven very fast by the strong electric field to reach the junction sides. In making the i-layer sufficiently thick for the photon absorption, mostly several microns, the efficiency may reach values of 0.8 or even above. A PIN-diode, typically fabricated using doped silicon, is show in Fig. 5.9b.

The photon absorption within the high field region of the junction gives not only higher efficiency but also faster response because of the absence of diffusion time. The maximum frequency response is then determined by the transit time through the junction. The carriers move at velocities that are limited by lattice scattering and are in the order of 10^6–$10^7\ \mathrm{cm\,s^{-1}}$. The transit times are therefore of the order of 10^{-11}–$10^{-10}\ \mathrm{s}$ corresponding to frequency ranges that may extend up to $10^{11}\ \mathrm{Hz}$. In practice the reachable frequencies are much lower because of parasitic inductance and capacitance, although for the usually small diode area the RC-time of the junction does not limit the frequency range. However, high frequency response in the order of $10^{11}\ \mathrm{Hz}$

has been reported [21] for a metal–semiconductor or Schottky photodiode consisting of semitransparent platinum and doped layers of GaAs. The high frequency response was reached by minimizing the parasitic circuit parameters of the diode construction.

5.2.4 Operational Modes

As we have discussed above the photon current is reverse. The current–voltage characteristic of the illuminated diode is therefore

$$i = i_d \left(e^{eV/kT} - 1 \right) - i_s, \tag{5.53}$$

where i_s is given by (2.3). The current–voltage characteristics with and without illumination are shown in Fig. 5.10.

As detector the photodiode can either operate in the *open circuit, current circuit*, or in the *reverse-biased circuit*. As derived from (5.53) the current in the *short circuit* for $V = 0$ is i_s and the voltage drop in the *open circuit* for $i = 0$ is

$$V = \frac{kT}{e} \ln \left(1 + \frac{i_s}{i_d} \right). \tag{5.54}$$

The *reverse-biased* mode (see Fig. 5.14) is mostly applied. Its operational behavior can be illustrated with Fig. 5.11.

The diode characteristics with and without illumination together with the straight load resistance line and the applied circuit voltage V_0 are shown. The applied voltage is divided over the diode and the load resistor. Without illumination the voltage drop over the load resistance is $V_0 - V_1$ and over the diode V_1. With illumination the load voltage changes to $V_0 - V_2$, whereas $V_1 - V_2 = V_s$ is the signal voltage. It is seen from the figure that V_s increases with R_L. In the open circuit mode the signal voltage is V_3 as indicated in Fig. 5.11. For sufficient large reverse voltage over the diode the current is saturated and the response is for small signals $V_2 - V_1 = i_s R_L$, whereas for the open circuit we find with (5.54) for $i_s \ll i_d$ that $V_s = \frac{kT}{ei_d} i_s = R_d i_s$ where R_d is the zero current diode resistance given by (5.41). The ratio of the responsivity R_L/R_d can be chosen much larger than one; hence the preference of the relatively simple reverse-biased mode for high response.

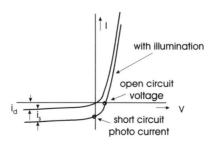

Fig. 5.10. Current–voltage characteristics of a photodiode with and without illumination

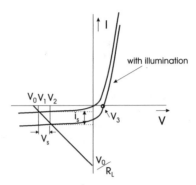

Fig. 5.11. A current–voltage diagram of the reverse biased photodiode circuit. The total applied voltage over the load resistance R_L in series with the photodiode is V_0. The distributions of voltage drops over load resistance and photodiode for both with and without illumination can be read from the diagram

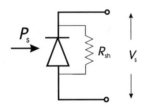

Fig. 5.12. Open circuit of a photodiode

5.2.5 Open Circuit

In the open circuit, indicated with Fig. 5.12, the photocurrent builds a potential over the space-charge region. The voltage is derived from (5.53) and given by (5.54). This build-up voltage on its turn generates a forward current which is equal to the reverse currents $i_s + i_d$ so that the net current is zero. The diode signal may, however, be reduced by a shunt resistor which is added for faster response or by the leakage resistance at the edges of the junction which may be due to the manufacture process. Including the shunt resistor R_{sh} we have for the open circuit the condition

$$-\frac{V}{R_{sh}} = i_d \left(e^{eV/kT} - 1 \right) - i_s .$$
(5.55)

For small signals with $eV \ll kT$ we derive from (5.55) by substituting (5.41)

$$V = \frac{R_d R_{sh}}{R_d + R_{sh}} i_s .$$
(5.56)

Substituting (2.3) we obtain for the responsivity $r = V/P_s$ of weak signals

$$r = \frac{R_d R_{sh}}{R_d + R_{sh}} \frac{\eta e}{h \nu_s} .$$
(5.57)

Because of the expansion made to derive (5.56) it should be noted that the open circuit mode is only linear for weak signals.

The noise currents associated with each of the two opposing current flows are not correlated so that by calculating the shot noise we take the sum of the two current flows. Including also the background current i_b the total reverse current is $i_d + i_s + i_b$. Assuming $R_{sh} \gg R_d$ this current is equal to the forward directed current $i_d e^{eV/kT}$. The shot noise is therefore

$$\overline{i_n^2} = 4eB \left(i_d + i_s + i_b \right) . \tag{5.58}$$

The dynamic diode resistance $R_d = \frac{dV}{di}$ at the voltage V is given by

$$\frac{1}{R_d} = \frac{d}{dV} \left[i_d \left(e^{eV/kT} - 1 \right) \right] = \frac{e i_d}{kT} e^{eV/kT} \tag{5.59}$$

or

$$\frac{1}{R_d} = \frac{e}{kT} \left(i_d + i_s + i_b \right) . \tag{5.60}$$

For the total noise we also have to add the Johnson noise of the shunt resistor. (The Johnson noise of the diode is included in the above calculated shot noise.)

The signal-to-noise ratio becomes by using (2.3), (2.18), and (2.22)

$$\frac{S}{N} = \frac{P_s^2}{\frac{4h\nu_s B}{\eta} \left[P_s + h\nu_s n + \frac{h\nu_s i_d}{e\eta} + \frac{h\nu_s k T_{eff}}{e^2 \eta R_{sh}} \right]} , \tag{5.61}$$

where the effective temperature T_{eff} of the shunt resistor includes the noise of the amplifier. Substituting (5.60) we obtain

$$\frac{S}{N} = \frac{P_s^2}{\frac{4h\nu_s B}{\eta} \left[P_s + h\nu_s n + \frac{h\nu_s i_d}{e\eta} \right] \left[1 + \frac{R_d T_{eff}}{R_{sh} T} \right]} , \tag{5.62}$$

where T is diode temperature.

If $\frac{R_d T_{eff}}{R_{sh} T} \ll 1$ which is usual in practice the NEP is dark current limited with

$$\text{NEP}_{DL} = \frac{2h\nu_s}{e\eta} \sqrt{e i_d B} . \tag{5.63}$$

The specific detectivity becomes with (2.2)

$$D^* = \frac{e\eta}{h\nu_s} \left[\frac{A}{4 e i_d} \right]^{1/2} . \tag{5.64}$$

Substituting the current responsivity $r_c = \frac{i_s}{P_s}$ we get

$$D^* = r_c \left[\frac{A}{4 e i_d} \right]^{1/2} . \tag{5.65}$$

It should be noted that for $R_{sh} \gg R_d$ the time constant of the detector is by using (5.41) equal to $C_d R_d = \frac{kT C_d}{e i_d}$. This value is mostly too large for high frequency response. Higher frequency operation and lower detectivity are obtained by adding a shunt resistance for which $R_{sh} < R_d$.

Example

A high detectivity PbSnTe photodiode [22] has in the open circuit a responsivity of $r = 3150\,\mathrm{V\,W^{-1}}$ for 10.6 μm radiation. The zero current diode resistance is $R_\mathrm{d} = 400\,\Omega$ and the diode area is $2.8 \times 10^{-3}\,\mathrm{cm^2}$. The measured D^* value was $10^{11}\,\mathrm{cm\,Hz^{1/2}\,W^{-1}}$.

Calculating the detectivity we note that for weak signals with $i_\mathrm{s} \ll i_\mathrm{d}$ we use (5.41) with $ei_\mathrm{d} = kT/R_\mathrm{d}$. From (5.65) and with $r_\mathrm{c} = r/R_\mathrm{d}$ we then obtain $D^* = 6.6 \times 10^{10}\,\mathrm{cm\,Hz^{1/2}\,W^{-1}}$, in substantial agreement with the measured value and close to the background limited ideal detectivity D_i^*.

5.2.6 Current Circuit

The photodiode can operate in the current mode by means of an operational amplifier which effectively hold the diode voltage at zero. The operational amplifier is discussed in Sect. 7.1. This scheme is depicted in Fig. 5.13.

The short circuit photocurrent is indicated on the illuminated characteristic in Fig. 5.10. The short circuit current can be measured by connecting the detector with the high impedance input terminals of the operational amplifier in such a way that the n-side of the diode goes to the $(-)$ terminal and the p-side to the $(+)$ terminal. The $(-)$ terminal of the output of the amplifier is connected via a feedback resistance R_f to the $(+)$ input terminal. The amplifier has a high gain, a high input impedance, and a low output impedance. The negative feedback drives the amplifier, depending on its gain, to a state where the voltage difference of the input terminals is at minimum, which is practically at zero voltage. Then the current through the feedback resistance R_f is equal to the signal current. Thus the output voltage of the operational amplifier is as discussed in Sect. 7.1

$$v_\mathrm{s} = R_\mathrm{f} i_\mathrm{s}\,, \tag{5.66}$$

which is, because of $V = 0$, not effected by any leakage resistance of the diode. Because R_f is much larger than the diode resistance R_d the output voltage is much higher than obtained with the open circuit mode.

At zero voltage the shot noise of the diode is

$$\overline{i_\mathrm{n}^2} = 2eB\left(2i_\mathrm{d} + i_\mathrm{s} + i_\mathrm{b}\right)\,. \tag{5.67}$$

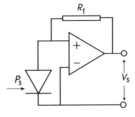

Fig. 5.13. Current circuit of a photodiode with an operational amplifier

Including the Johnson noise of the shunt resistance R_{sh} of the diode, which is added for faster response or is associated with the current leakage at the edges of the junction region, we obtain for the signal-to-noise ratio

$$\frac{S}{N} = \frac{P_s^2}{\frac{2h\nu_s B}{\eta}\left[P_s + h\nu_s n + \frac{2h\nu_s i_d}{e\eta} + \frac{2h\nu_s kT_{eff}}{e^2\eta R_{sh}}\right]} . \tag{5.68}$$

If we are dealing with the common situation that the background noise is negligible and if we use (5.41) and assume as usual in practice, $R_{sh}T > R_dT_{eff}$ the detector is dark current limited. We obtain from (5.68)

$$\mathrm{NEP_{DL}} = \frac{2h\nu_s}{e\eta}\sqrt{ei_d B} \tag{5.69}$$

and with (2.2)

$$D^* = \frac{e\eta}{2h\nu_s}\sqrt{\frac{A}{ei_d}} , \tag{5.70}$$

which is the same as obtained for the open circuit. The corresponding response time is

$$\tau = \frac{kTC_d}{ei_d} , \tag{5.71}$$

where we have substituted (5.41).

Higher frequency response is obtained for $R_{sh} < R_d$. The detector becomes amplifier limited and we obtain from (5.68)

$$\mathrm{NEP_{AL}} = \frac{2h\nu_s}{e\eta}\sqrt{\frac{kT_{eff}B}{R_v}} \tag{5.72}$$

and

$$D^* = \frac{e\eta}{2h\nu_s}\sqrt{\frac{AR_v}{kT_{eff}}} \tag{5.73}$$

with the response time

$$\tau = R_v C_d , \tag{5.74}$$

where $R_v = \frac{R_d R_{sh}}{R_d + R_{sh}}$.

5.2.7 Reverse-Biased Circuit

In the reverse-biased mode operation is most common because of its simplicity and the much higher response than in the case of the open circuit mode. The operation is depicted in Fig. 5.14.

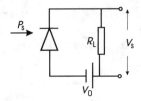

Fig. 5.14. Circuit of a reverse-biased photodiode

The diode resistance becomes very large for the applied voltage $|V| \gg \frac{kT}{e}$ as can be seen from Fig. 5.10 so that the generated signal current flows through the load resistance and the shunt resistance of the diode but practically not through the diode junction. The current through the load resistor is then

$$i_1 = \frac{R_{\text{sh}}}{R_{\text{sh}} + R_{\text{L}}} i_{\text{s}} . \tag{5.75}$$

Eliminating the constant i_{d} current by operating the signal in the ac mode and applying a blocking capacitor, the voltage response $r = \frac{V_{\text{s}}}{P_{\text{s}}}$ becomes

$$r = \frac{R_{\text{L}} R_{\text{sh}}}{R_{\text{L}} + R_{\text{sh}}} \frac{e\eta}{h\nu_{\text{s}}} . \tag{5.76}$$

Comparing this last result with (5.57) it is seen that by taking R_{L} much larger than R_{d} the response of the reverse-biased mode is much larger than that of the open circuit mode. However, it does not make sense to increase R_{L} if $R_{\text{L}} \gg R_{\text{sh}}$. It may happen that the resistance is determined by the desired response time of the detector which is given by

$$\tau = \frac{R_{\text{L}} R_{\text{sh}}}{R_{\text{L}} + R_{\text{sh}}} C_{\text{d}} , \tag{5.77}$$

where C_{d} is given by (5.25).

Considering the noise we note that the forward current is suppressed so that the shot noise is only produced by the reverse current and is given by

$$\overline{i_{\text{n}}^2} = 2eB(i_{\text{d}} + i_{\text{s}} + i_{\text{d}}) \tag{5.78}$$

with a factor 2 smaller than in the open circuit. Looking at (5.61) the signal-to-noise ratio of the reverse-biased circuit is then given by

$$\frac{S}{N} = \frac{P_{\text{s}}^2}{\frac{2h\nu_{\text{s}}B}{\eta} \left[P_{\text{s}} + h\nu_{\text{s}}n + \frac{h\nu_{\text{s}}i_{\text{d}}}{e\eta} + \frac{2h\nu_{\text{s}}kT_{\text{eff}}}{e^2\eta R_v} \right]} , \tag{5.79}$$

where $R_v = R_{\text{L}} R_{\text{sh}}/(R_{\text{L}} + R_{\text{sh}})$. In case R_{sh} is much larger than R_{L} the value of R_v becomes practically equal to R_{L}.

The S/N-ratio can be increased by minimizing the Johnson noise, which is obtained for $R_v T \gg 2R_d T_{\text{eff}}$. Then the minimum NEP is dark current limited and given by

$$\text{NEP}_{\text{DL}} = \frac{h\nu}{e\eta}\sqrt{2ei_d B} \tag{5.80}$$

and with (2.2) the detectivity by

$$D^*_{\text{DL}} = \frac{e\eta}{h\nu}\sqrt{\frac{A}{2ei_d}}, \tag{5.81}$$

which is a factor $\sqrt{2}$ larger than in the open circuit and current circuit modes.

Since i_d according to (5.40) is proportional to A we find that D^*_{DL} is only determined by material properties and is *independent* on its dimensions and reverse biased voltage. The values of D^*_{DL} are plotted in Fig. 5.5 as a function of wavelength for various photodiodes.

We remark that a large value of R_v gives rise to a large response time of the detector which may not be desirable. A compromise is usually made by choosing a value of R_v for which the Johnson noise becomes equal to the dark current noise or $R_v = \frac{2kT_{\text{eff}}}{ei_d}$.

Dealing with the common situation that the background noise is negligible relative to the dark current noise the noise-equivalent-power is then derived by substituting $R_v = \frac{2kT_{\text{eff}}}{ei_d}$ into (5.79). We obtain

$$\text{NEP} = \frac{2h\nu_s}{e\eta}\sqrt{ei_d B}. \tag{5.82}$$

In this way the NEP and D^* are equal to those obtained for the open and current circuits. However, the corresponding response time is $\tau = R_v C_d$ or

$$\tau = \frac{2kT_{\text{eff}}C_d}{ei_d}, \tag{5.83}$$

which looks at first glance a factor 2 larger than in the previous cases. However, the capacitance of the reverse-biased diode is smaller and decreases with the applied voltage as given by (5.24) and (5.25). Thus the frequency response depends on the bias voltage.

Example

A commercial germanium photodiode, reverse biased at 10 V, has at room temperature for 1.06 µm radiation the following specifications: quantum efficiency 73%, current response 0.7 A W^{-1}, dark current 6 µA, wide band noise current 2×10^{-12} A Hz$^{-1/2}$, NEP $= 3 \times 10^{-12}$ W Hz$^{-1/2}$, capacitance 28 pF and response time 250 ns.

We substitute the current response $r_c = \frac{e\eta}{h\nu} = 0.7$ and the noise current $\sqrt{4ei_d} = 2 \times 10^{-12}$ A Hz$^{-1/2}$ into (5.82) and find NEP $= 2.8 \times 10^{-12}$ W Hz$^{-1/2}$.

The time response obtained with (5.83) gives $\tau = 2.3 \times 10^{-7}$. Both NEP and τ are in good agreement with the quoted experimental data.

Faster response is obtained for $R_v < \frac{2kT_{\text{eff}}}{ei_d}$. However, the faster the system the more noisy and the smaller the response. In that case the Johnson noise is dominant and the detector is amplifier limited. We obtain from (5.79)

$$\text{NEP}_{\text{AL}} = \frac{2h\nu_s}{e\eta} \sqrt{\frac{kT_{\text{eff}}B}{R_v}} \tag{5.84}$$

or with (2.2)

$$D^* = \frac{e\eta}{2h\nu_s} \sqrt{\frac{AR_v}{kT_{\text{eff}}}}. \tag{5.85}$$

Finally we remark that also the reverse-biased mode is often applied in combination with an operational amplifier as discussed in Sect. 7.1.

5.3 Avalanche Photodiodes

At increasing reverse bias voltage the photodiode enters the avalanche region before it breaks down at large reverse-biased voltages. The increasing field reaches the point at which carriers that are accelerated across the space-charge region gain enough kinetic energy to excite electrons from the valence band into the conduction band, thus producing additional electron–hole pairs. These newly generated electrons and holes drift in turn in opposite directions. On their way they too may gain sufficient energy to produce electron–hole pairs. Thus holes and electrons can both cause a multiplication of electron–hole pairs. Because the carriers drift in opposite directions toward the boundaries of the space-charge region the current in the external circuit is then also multiplied by the same factor. The production of electron–hole pairs is described by the ionization coefficients of electrons and holes. They are strong functions of the electric field. Since the resistance of the space-charge region is much larger than that of the bulk material the applied voltage is across the space-charge region where the ionization occurs. To attain high performance with low noise and fast response the ratio of the ionization coefficients of electrons and holes should be very different from unity as will be discussed.

The avalanche photodiodes with their internal gain combine the benefits of both PIN photodiodes and photomultipliers. For many applications like light wave communication systems these high-gain avalanche photodiodes offer considerable advantages over normal photodiodes. This is especially the case if fast diode detectors with small time constants and thus with high Johnson noise have to be used. The avalanche process reduces the relative contribution of the Johnson noise in a similar way as in the photomultiplier process.

For instance in the wave length region of 0.8–0.9 μm the silicon avalanche photodiode is an attractive detector with respect to its relatively low noise and fast response. Longer wavelength systems with low noise have been developed by using combinations of binary (InP,GaSb) and ternary/quaternary (InGaAs, InGaAsP, AlGaAsSb) III–V semiconductors. Unfortunately, most other combinations of III–V semiconductors have comparable ionization coefficients for electrons and holes, which have a very unfavorable effect on the noise generated in the multiplication process.

The construction of the avalanche photodiode is similar to normal photodiodes, except that for obtaining uniform amplification over the surface special attention is paid to the uniformity of the junction. Very successful is the development of the silicon PIN photodiode (see Fig. 5.9b) with its quantum efficiency as high as 90% throughout the visible spectrum and with no diffusion time (fast response). Excellent doping uniformity and a low number of lattice defects enable the fabrication of large devices up to 20 mm diameter with gains of several hundred and breakdown voltages of 2 kV and higher. The construction is usually such that the entrance p-layer is very thin, less than 1 μm, so that the incident power is absorbed in the intrinsic region, where the avalanche gain is build-up. For silicon the ionization coefficient of electrons is much larger than that of holes so that the free carriers are mainly produced by the electrons due to the high reverse-bias voltage creating a strong field in the intrinsic region.

5.3.1 Multiplication Process

In Fig. 5.15 a scheme for the avalanche multiplication is shown. The spatial variation of the electric field is arbitrary but in the direction of the electric field. The space-charge region has a width w. As shown the electrons drift in the positive direction with velocity v_n and the holes in the negative direction with velocity v_p. The current density for electrons J_n and holes J_p are related to the carrier densities n and p by $J_n = -env_n$ and $J_p = epv_p$ where e is positive and equal to the absolute value of the electron charge. The total current $J = J_n + J_p$ is positive in the direction of the field, but in the multiplication process $|J_n|$ increases, whereas $|J_p|$ decreases with increasing x.

The electric field of the applied voltage is in the space-charge region because of its high resistance. The electron–hole-pair creation occurs at fields of the order of 10^5 V cm^{-1} and is described by the ionization coefficients α and

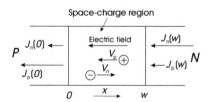

Fig. 5.15. Scheme of the avalanche multiplication process of the photodiode. The space charge region has a width w. The electric field, current densities of electrons and holes, and their velocities are indicated

β for electrons and holes, respectively. The general differential equations for the avalanche currents under quasi-steady-state conditions are

$$\frac{\mathrm{d}J_n\left(x\right)}{\mathrm{d}x} = \alpha J_n\left(x\right) + \beta J_p\left(x\right) \tag{5.86}$$

and

$$-\frac{\mathrm{d}J_p\left(x\right)}{\mathrm{d}x} = \alpha J_n\left(x\right) + \beta J_p\left(x\right) , \tag{5.87}$$

where α and β depend on the local field and are therefore functions of x. It is seen that $\mathrm{d}J_n(x)/\mathrm{d}x + \mathrm{d}J_p(x)/\mathrm{d}x = 0$ so that for the steady state the total current is constant.

$$J_n\left(x\right) + J_p\left(x\right) = J = \mathrm{const.} \tag{5.88}$$

The photon current avalanche gain M is defined as the ratio of the current flowing through the diode in the presence of the multiplication gain to the current in the absence of gain. In the case the initial photon ionization occurs in the p region so that the corresponding photon current of the minority carriers is $J_n(0)$ the avalanche gain is

$$M_n = \frac{J}{J_n\left(0\right)} . \tag{5.89}$$

Similarly, if the photon absorption occurs in the n-region with the corresponding photon current $J_p(w)$ of the hole minority carriers the avalanche gain is

$$M_p = \frac{J}{J_p(w)} . \tag{5.90}$$

The solution of the avalanche rate equations (5.86) and (5.87), shown in Appendix A.3, yields for the gain factors

$$M_n = \left\{1 - \int_0^w \alpha \exp\left[-\int_0^x \left(\alpha - \beta\right) \mathrm{d}x'\right] \mathrm{d}x\right\}^{-1} \tag{5.91}$$

and

$$M_p = \left\{1 - \int_0^w \beta \exp\left[-\int_0^x \left(\alpha - \beta\right) \mathrm{d}x'\right] \mathrm{d}x\right\}^{-1} . \tag{5.92}$$

The avalanche breakdown voltage is the voltage at which the gain goes to infinity. This happens when the denominator of (5.91) or (5.92) becomes zero. However, for the special case $\beta = 0$ we derive from (5.86)

$$M_n\left(\beta = 0\right) = \exp\left[\int_0^w \alpha \,\mathrm{d}x\right] , \tag{5.93}$$

which shows that breakdown does not occur under this condition and that the current remains always finite.

5.3.2 Multiplication Noise

The noise at the output of the avalanche photodiode does not only come from the amplified signal input noise but also from the multiplication process itself. The noise generated in the multiplication process depends critically on the relative magnitudes of the rates α and β [23]. In general, the most favorable condition is when one of the ionization coefficients is zero and the worst case is found for $\alpha = \beta$. It is instructive and relatively simple, compared with the general situation, to treat the noise problem for these two extreme conditions. Doing this we consider for the sake of simplicity the initial photon absorption to take place only in the p-region with a photon current $J_n(0)$ entering the space-charge, see Fig. 5.15. Besides, by calculating the noise of the multiplication process we take advantage of the similarity with the photomultiplier treated in Sect. 4.2.

Case 1, $\beta = 0$

The amplified input noise is equal to $2eJ_n(0)M_n^2B$. Similar to the generated current at each stage of the photomultiplier we also assume for the avalanche photodiode that the increase of current at any position x in the space-charge region obeys Poisson statistics. Thus the current increase $dJ_n(x)$ generates shot noise equal to $2eB\,dJ_n(x)$. This shot noise current will then be amplified by the electric field over the remaining distance through the space-charge region. Since the current at the position x is equal to $J_n(x)$ the remaining gain from the position x is $M_n(x) = J/J_n(x)$ so that any fluctuation of the current $J_n(x)$ will be further amplified by $M_n(x)$. Integrating all these amplified noise currents we obtain for the total shot noise

$$\overline{i_n^2} = 2eB\left[M_n^2 J_n(0) + \int_{J_n(0)}^{J} M_n^2(x)\,dJ_n(x) \right]. \tag{5.94}$$

Substituting $M_n(x) = J/J_n(x)$ and $J = M_n J_n(0)$ (5.94) yields

$$\overline{i_n^2} = 2eBJ_n(0)M_n^2\left(2 - \frac{1}{M_n} \right). \tag{5.95}$$

Thus, the shot noise for large values of M_n is twice the value obtained with the ideal multiplier.

Case 2, $\alpha = \beta$

When both carriers have the same ionization coefficient the treatment is more complicated. For $\alpha = \beta$ (5.86) becomes

$$\frac{dJ_n(x)}{dx} = \alpha J \tag{5.96}$$

with the solution

$$J_n(w) = J \int_0^w \alpha \, dx + J_n(0) \, . \tag{5.97}$$

Let us now look at the amplification of an incremental increase of current $\Delta J_n(x)$ through the space-charge region. This current element will be subsequently amplified and reach the end for $x = w$ at the right of the space-charge, shown in Fig. 5.15. According to (5.96) the amplified current $\Delta J_n(w)$ of this element will be

$$\Delta J_{n,r}(w) = J \int_x^w \alpha \, dx + \Delta J_n(x) \, . \tag{5.98}$$

However, since in the amplification process equal amounts of holes and electrons are generated there is simultaneously a hole current $\Delta J_p(x)$ equal to $\Delta J_n(x)$ flowing in the opposite direction to the other end of the space-charge. This current part will then also be amplified and reaches at $x = 0$ the value

$$\Delta J_{n,l}(w) = -J \int_x^0 \alpha \, dx \, . \tag{5.99}$$

The total amplification experienced by a current increase $\Delta J_n(x)$ is then the sum of the (5.98) and (5.99) or

$$\Delta J_n(w) = J \int_0^w \alpha \, dx + \Delta J_n(x) \, . \tag{5.100}$$

If we now compare (5.100) with (5.97) we find that the gain of the current increase is equal to that of the current input. In other words a small increase in current anywhere in the space-charge region will be amplified in the space-charge region by the same gain factor M_n. Applying this result in (5.94) yields

$$\overline{i_n^2} = 2eB \left[M_n^2 J_n(0) + M_n^2 \{ J - J_n(0) \} \right] = 2eB J_n(0) M_n^3 \, . \tag{5.101}$$

Since for an ideal multiplier the amplified shot noise goes with M_n^2 it is customary to write the ratio of the output to the input shot noise of the avalanche photodiode as $M_n^2 F$ where F is the avalanche or excess-noise factor. Thus for $\beta = 0$ we found $F = 2$ and for $\alpha = \beta$, $F = M_n$. A comprehensive analysis has shown that F strongly depends on the ratio of the ionization coefficients [23]. The lowest excess-noise factor is obtained when α/β is either very large or very small and when the multiplication process is initiated by the carrier with the highest ionization coefficient. Experimentally it has been found that the output shot noise of high performance avalanche photodiodes behaves more or less as $M_n^{2.1}$ so that $F = M_n^{0.1}$.

5.3.3 Detectivity

The electric circuit with reverse-biased voltage is shown in Fig. 5.14. Since the avalanche photodiode is the solid state analog to the vacuum photomultiplier discussed in Sect. 4.2 the signal-to-noise ratio is similar to the expression (4.20). We then write

$$\frac{S}{N} = \frac{P_s^2}{\dfrac{2h\nu_s FB}{\eta}\left[P_s + h\nu_s n + \dfrac{h\nu_s i_d}{e\eta} + \dfrac{2h\nu_s kT_{\text{eff}}}{FM_n^2 e^2 \eta R_v}\right]},\tag{5.102}$$

where we assume that the dark current undergoes the same gain as the signal and $R_v = R_L R_{\text{sh}}/(R_L + R_{\text{sh}})$. The amplifier noise term of the ordinary reverse-biased photodiode equal to $4kT_{\text{eff}}B/R_v$ is typically much larger than the shot noise terms when dealing with fast response. It is seen that in the case of amplifier limitation the $\frac{S}{N}$-value increases with M_n^2. This improvement continues until the shot noise terms become comparable with the amplifier noise. If M_n is sufficiently high to neglect the amplifier noise and we are dealing with negligible background noise compared with the dark current noise the minimum NEP becomes

$$\text{NEP} = \frac{h\nu_s}{\eta}\left(\frac{2F i_d B}{e}\right)^{1/2}.\tag{5.103}$$

The specific detectivity becomes according to (2.2)

$$D^* = \frac{\eta}{h\nu_s}\left(\frac{eA}{2F i_d}\right)^{1/2}.\tag{5.104}$$

If F can be approximated by $M_n^{0.1}$ its effect on NEP and D^* is negligible. According to (5.40) the saturation current i_d is proportional to A so that the detectivity is only determined by the material properties of the diode. However, this is not the case if the dependence of F on M_n is strong like $F = M_n$ found for $\alpha = \beta$. In that case the minimum NEP is reached for M_n equal to twice the ratio of the amplifier noise to the dark current shot noise in the absence of amplification. (The background noise is neglected.) Further increase of M_n results in an increase of NEP. Thus for $\alpha = \beta$ the minimum NEP becomes

$$\text{NEP} = \frac{h\nu_s}{\eta}\left[\frac{8kT_{\text{eff}}B}{e^2 R_v} + \frac{R_v i_d^2 B}{4kT_{\text{eff}}}\right]^{1/2}\tag{5.105}$$

and maximum specific detectivity

$$D^* = \frac{\eta}{h\nu_s}\left[\frac{8kT_{\text{eff}}}{e^2 R_v A} + \frac{R_v i_d^2}{4kT_{\text{eff}} A}\right]^{-1/2}.\tag{5.106}$$

It should be noted that cooling the avalanche photodiode reduces the dark current of the thermally generated carriers (and hence the dark current noise). It also lowers the breakdown voltage. Although the development of the avalanche photodiodes is impressive, their gain and NEPs are still inferior compared with those of the vacuum photomultipliers. For instance the photomultiplier can count single photons under optimum conditions, whereas avalanche photodiodes have equivalent noise charges corresponding to roughly 25 electrons. Further developments are still challenged to bridge this gap. Substantial progress for obtaining higher gain and lower noise was based on multiple junctions and the enhancement of the α/β-ratio. Of interest are multiquantum well avalanche photodiodes [24], superlattice avalanche photodiode with periodic doping profile [25] and superlattice avalanche photodiode with graded gap sections [26]. A detailed review is found in [27].

5.3.4 Frequency Response

By considering the frequency response of avalanche photodiodes we assume that the initial photon absorption occurs at one edge of the space-charge region so that diffusion into the space-charge is absent. Then there are three time constants involved in determining the frequency response: firstly, the space-charge region transit time; secondly, the avalanche build-up time; thirdly, the RC-time constant of the device. These time constants depend strongly on the semiconductor material, the size of the junction area, and the length of the space-charge region. The latter is about the inverse value of the light absorption coefficient of the material. These time constants may differ considerably for various systems. For instance the light absorption coefficient in the wavelength region of interest is for germanium an order of magnitude larger than for silicon so that the length of the space-charge in germanium is chosen an order of magnitude smaller than that for silicon. In a silicon avalanche photodiodes the space-charge region needs to be at least 30–50 μm long for good quantum efficiency and thereby the transit time governs the response speed. In germanium, however, the transit time is not the limiting factor because the space-charge is not more than 2 or 3 μm long, which results for the saturated drift velocity $(6 \times 10^6 \text{ cm s}^{-1})$ to a transition time of about 5×10^{-11} s. Therefore with germanium the RC-time or the avalanche build-up time governs the response time.

The RC-time is given by

$$\tau_{RC} = R_v C_d , \tag{5.107}$$

where $R_v = R_L R_{sh}/(R_L + R_{sh})$ and C_d is the sum of the junction capacitance given by (5.25) and parasitic capacitance due to construction parameters. In the reverse-biased mode the large resistance of the diode itself has no effect on the time constant. For a small time constant the diode area should be made as small as possible. For example having a diode capacitance of 3 pF for germanium and $R_v = 50 \, \Omega$ we have $\tau_{RC} = 1.5 \times 10^{-10}$ s.

The avalanche multiplication builds up by contributions of carrier feedback. Numerical studies have shown [28] that the build-up time depends on the number of feedback processes. For $\alpha = \beta$ the transit time is proportional to M_n because during the effective build-up time of the multiplication process there are M_n transits across the space-charge. However, the effective response time depends strongly on the ratio of α and β. Computer studies have shown that the avalanche multiplication does not reduce the frequency response as long as M_n is less than the ratio of α and β [28]. For instance, germanium having typically a small length of the space-charge region (about $3\,\mu$m) is nevertheless not the ideal material for avalanche photodiodes because of its nearly equal ionization coefficients for electrons and holes. In contrast, silicon with its large α/β-value and a transit time that is an order of magnitude larger than that of germanium has nevertheless an effective build-up time that is an order of magnitude smaller.

Fortunately, the condition for shortening the avalanche build-up time also minimizes the multiplication noise.

6

Correlation Analyses

Correlation functions and their mathematical manipulations are extremely useful for recovering information buried in noise. For example, a periodic signal mixed with noise can be selectively filtered by a correlation process, whereas the accompanying noise is continuously suppressed so that in due time a clean signal can be provided. Even, if a stationary physical process produces several simultaneous signals, each of them is any variation with time, a complete description of these signals and their correlation shifts can be provided by the analysis of the correlation waveform.

The considered signal processing is either based on autocorrelation or on cross correlation. Autocorrelation can be used to recover unknown repetitive signals, or to measure a particular band of signals or noise frequencies, whereas cross correlations are applied when the signal frequency is known and the waveform itself has to be investigated. This chapter deals with theoretical background whereas in Chap. 7 we shall discuss instrumentation based on correlation processes.

6.1 AutoCorrelation

Autocorrelation involves the integration of the product of a waveform containing a current signal of some stationary process and its accompanying noise with a delayed function of itself. The usual mathematical expression is

$$\phi_{\text{AA}}\left(\tau\right) = \lim_{T \to \infty} \frac{1}{2T} \int_{-T}^{T} f_{\text{A}}\left(t + \tau\right) f_{\text{A}}\left(t\right) \mathrm{d}t , \qquad (6.1)$$

where the function $f_{\text{A}}(t)$ describes the signal mixed with noise. The autocorrelation provides a measure of the "memory" of the process. Therefore, it contains information on the stationary process and rejects the random noise. It is seen that the autocorrelation is an even function of the time delay τ, and that for $\tau = 0$ its value is the mean square of the process. When the integration of (6.1) is performed for many values of τ, the complete correlation function is obtained. The autocorrelation extracts the signal buried in noise but rejects phase information of periodic components of the signal. Therefore, it produces a waveform that may not be an actual reproduction of the input waveform. It will, however, always have the same periodicities as the input signal. This can be demonstrated by considering the autocorrelation of a square wave which produces a triangular output function as shown in Fig. 6.1.

For spectral analysis, it is useful to convert the autocorrelation function, described in the time domain τ, into the power density spectrum in the frequency domain. The Fourier transform $F_A(\omega)$ of $f_A(t)$ is given by (1.16). If we substitute the Fourier transform of $f_A(t)$ into the integrand of (6.1) and interchange the order of integration we obtain

$$\lim_{T \to \infty} \frac{1}{2T} \int_{-T}^{T} f_A(t+\tau) f_A(t) \, dt = \lim_{T \to \infty} \frac{1}{2\pi T} \int_{0}^{\infty} |F_A(\omega)|^2 \cos \omega\tau \, d\omega . \quad (6.2)$$

Although integration over T goes to infinity, (6.2) can also be considered for a finite time by setting $f_A(t)$ to zero outside the integration interval. In that case the function $\phi_{AA}(\tau)$ is called the covariance. If, however, the process is stationary i.e., the averages of $f_A(t)$ and $f_A^2(t)$ are independent of the time interval at which they are computed as is for instance, the case with white noise currents, the covariance function would be independent on time and hence equal to the autocorrelation function.

Suppose that $f_A(t)$ is the current over a resistor of $1\,\Omega$ and that the process is observed over the time interval $[T, -T]$ so that outside this interval $f_A(t)$ can be regarded as zero. Then, when $\tau = 0$, (6.2) reduces to the average power P over the period $2T$ or

$$P = \frac{1}{2T} \int_{-T}^{T} [f_A(t)]^2 dt = \frac{1}{2\pi T} \int_{0}^{\infty} |F_A(\omega)|^2 \, d\omega . \quad (6.3)$$

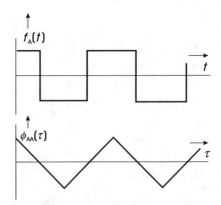

Fig. 6.1. Autocorrelation of a square wave

The spectral power density $S_{AA}(\omega) = dP/d\omega$ becomes

$$S_{AA}(\omega) = \frac{1}{2\pi T}|F_A(\omega)|^2 \, . \tag{6.4}$$

The spectral power density is uniquely related to the autocorrelation of the process. Substituting $S_{AA}(\omega)$ into (6.2) we get

$$\phi_{AA}(\tau) = \int_0^\infty S_{AA}(\omega)\cos\omega\tau \, d\omega \, . \tag{6.5}$$

The Fourier transform of $\phi_{AA}(\tau)$ can be written directly according to (1.16).

$$S_{AA}(\omega) = \frac{2}{\pi}\int_0^\infty \phi_{AA}(\tau)\cos\omega\tau \, d\tau \, . \tag{6.6}$$

Thus we conclude that the frequency spectra of unknown repetitive signals, even buried in the accompanying noise, can be resolved by the Fourier transform of the autocorrelation. We shall see that the accompanying random noise decreases with increasing correlation time so that clean signal information is obtained after a sufficient long correlation time. Equations(6.5) and (6.6) are known as the Wiener-Khintchine theorem .

6.2 Cross Correlation

The cross correlation in analogy with the autocorrelation involves the integration of the product of two signals from different, but coherent sources. The resulting correlation function contains information regarding the frequencies that are common to both signals and the phase difference between them. The usual mathematical expression is

$$\phi_{AB}(\tau) = \lim_{T\to\infty} \frac{1}{2T}\int_{-T}^T f_A(t+\tau)f_B(t)\,dt \, , \tag{6.7}$$

where τ is a delay time. $f_A(t)$ and $f_B(t)$ are two different time functions that arise from the process being investigated. It follows from (6.7) that $\phi_{AB}(\tau) = \phi_{BA}(-\tau)$. As an example we show in Fig. 6.2, a noisy sine wave correlated against a noisy square wave of the same frequency, producing a sinusoidal correlation function that shows also the phase angle between the two correlated signals.

The cross correlation describes the degree of conformity between two different signals as a function of their mutual delay. The measured quantitative degree of likeness of the two signals can supply more insight into the phenomena of interest than the separate analysis of either signal alone. In practice, often one of the two signals is well defined and without noise and the cross correlation is carried out for eliminating the noise of the signal being investigated.

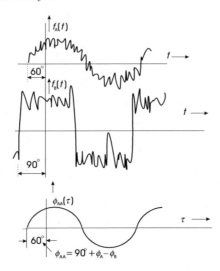

Fig. 6.2. Cross correlation of a noisy sine wave with a noisy square wave

Similar to the derivation of the spectral power density the cross spectral density is obtained by taking the Fourier transform of the cross correlation. Substituting the inverse formula for $f_A(t)$ as given by (1.17) in the integrand of (6.7) and change the order of integration we obtain

$$\lim_{T \to \infty} \frac{1}{2T} \int_{-T}^{T} f_A(t+\tau) f_B(t) \, dt = \lim_{T \to \infty} \frac{1}{4\pi T} \int_{-\infty}^{\infty} F_A(\omega) F_B^*(\omega) e^{j\omega\tau} d\omega ,$$

(6.8)

where $F_{A,B}$ are the Fourier transforms of $f_{A,B}$.

Observing the cross correlation over the interval $[-T, T]$ we can still apply (6.8) by considering outside this time interval either f_A or f_B equal to zero. For $\tau = 0$ (6.8) describes the cross power P over the period or

$$P = \frac{1}{2T} \int_{-T}^{T} f_A(t) f_B(t) \, dt = \frac{1}{4\pi T} \int_{-\infty}^{\infty} F_A(\omega) F_B^*(\omega) \, d\omega .$$

(6.9)

The cross spectral power density in analogy with $S_{AA}(\omega)$ is defined as

$$S_{AB}(\omega) = \frac{1}{2\pi T} F_A(\omega) F_B^*(\omega) .$$

(6.10)

Because F_A and F_B are in general complex $S_{AB}(\omega)$ is complex. It is seen from (6.10) that the following relations exist.

$$S_{AB}(\omega) = S_{AB}^*(-\omega) = S_{BA}(-\omega) = S_{BA}^*(\omega) .$$

(6.11)

Equation (6.8) can be written as

$$\phi_{AB}(\tau) = \frac{1}{2} \int_{-\infty}^{\infty} S_{AB}(\omega)\,e^{j\omega\tau}\,d\omega\,. \tag{6.12}$$

It is seen that $S_{AB}(\omega)$ can be considered as the Fourier transform of $\phi_{AB}(\tau)$ which is according to (1.16)

$$S_{AB}(\omega) = \frac{1}{\pi} \int_{-\infty}^{\infty} \phi_{AB}(\tau)\,e^{-j\omega\tau}\,d\tau\,. \tag{6.13}$$

6.2.1 Signal Recovery by Cross Correlation

We consider a weak periodic current signal, $f_S = i_s \cos\omega_s t$, that is very small in comparison to its accompanying background noise. This signal will be recovered in a cross correlation with a chosen reference signal $f_B = i_b \cos\omega_s t$. The noise bandwidth is outside the region of any disturbing periodic pickup signals, such as harmonics of power supplies or broadcasting systems and flicker noise. The accompanying noise current $f_N(t)$ is only white with constant spectral power density. In case of Johnson noise the spectral power density, $\overline{i_0^2}(\omega)$, in terms of radial frequency is equal to $4kT/2R\pi$ and in case of shot noise equal to ei_0/π. The signal observing time of the correlation process is $2T$ which corresponds to a bandwidth $\Delta\nu = 1/4T$ or $\Delta\omega = \pi/2T$. The noise current is then

$$\overline{i_n^2} = \overline{i_0^2}(\omega)\,\Delta\omega = \overline{i_0^2}(\omega)\,\frac{\pi}{2T}\,. \tag{6.14}$$

Applying (6.4) we derive for the Fourier transform of the noise current $F_N(\omega)$

$$|F_N(\omega)|^2 = 2\pi T \overline{i_0^2}(\omega)\,. \tag{6.15}$$

The Fourier transform of the periodic signal becomes by applying (1.16)

$$F_S(\omega) = 2i_s \int_0^T \cos\omega_s t \cos\omega t\,dt = i_s \left[\frac{\sin(\omega_s + \omega)T}{\omega_s + \omega} + \frac{\sin(\omega_s - \omega)T}{\omega_s - \omega} \right]\,. \tag{6.16}$$

For the reference signal we find similarly[1]

$$F_B = 2i_b \int_0^{\infty} \cos\omega_s t \cos\omega t\,dt = \pi i_b\,[\delta(\omega_s + \omega) + \delta(\omega_s - \omega)]\,. \tag{6.17}$$

---

[1] The δ-function $\delta(\omega_s - \omega)$ is defined by $\lim_{T\to\infty}[\sin(\omega_s - \omega)T/\pi(\omega_s - \omega)] = \delta(\omega_s - \omega)$

Substituting into (6.7), $f_A = f_S + f_N$, we get

$$\phi_{AB}(\tau) = \phi_{SB}(\tau) + \phi_{NB}(\tau) \, , \tag{6.18}$$

where

$$\phi_{SB}(\tau) = \frac{1}{2T} \int_{-T}^{T} f_S(t + \tau) f_B(t) \, dt \tag{6.19}$$

and

$$\phi_{NB}(\tau) = \frac{1}{2T} \int_{-T}^{T} f_N(t + \tau) f_B(t) \, dt \, . \tag{6.20}$$

Applying (6.8) we obtain for (6.19)

$$\phi_{SB}(\tau) = \frac{1}{4\pi T} \int_{-\infty}^{\infty} F_S(\omega) F_B e^{j\omega\tau} \, d\omega \, . \tag{6.21}$$

Substituting (6.16) and (6.17) into (6.21) we get

$$\phi_{SB}(\tau) = \frac{1}{2\pi T} \int_{0}^{\infty} \pi i_b \delta(\omega_s - \omega) \, i_s \frac{\sin(\omega_s - \omega)T}{\omega_s - \omega} \cos \omega\tau \, d\omega \tag{6.22}$$

or

$$\phi_{SB}(\tau) = \frac{1}{2} i_s i_b \cos \omega_s\tau \, . \tag{6.23}$$

For the accompanying white noise, the probabilities associated with the current fluctuations are invariant under a shift of time. So we find for (6.20) by applying (6.8)

$$\phi_{NB} = \frac{1}{4\pi T} \int_{-\infty}^{\infty} F_N(\omega) F_B d\tau \tag{6.24}$$

and by substituting (6.15) and (6.17); we find

$$\phi_{NB} = \frac{1}{2\pi T} \int_{0}^{\infty} \pi i_b \delta(\omega_s - \omega) \left[2\pi T \overline{i_0^2}(\omega)\right]^{1/2} d\omega = i_b \left[\frac{\pi \overline{i_0^2}(\omega)}{2T}\right]^{1/2} \tag{6.25}$$

or in terms of Hertz frequencies

$$\phi_{NB} = i_b \left[\frac{\overline{i_0^2}(\nu)}{4T}\right]^{1/2} . \tag{6.26}$$

Finally by substituting (6.23) and (6.26) into (6.18) we get

$$\phi_{AB}(\tau) = \frac{1}{2} i_s i_b \cos \omega_s\tau + i_b \left[\frac{\overline{i_0^2}(\nu)}{4T}\right]^{1/2} . \tag{6.27}$$

It is seen that the cross correlation signal contains a periodic term with amplitude $\frac{1}{2}i_s i_b$ and a noise term independent on τ. Taking the signal-to-noise ratio of the signal and noise powers we have

$$\frac{S}{N} = \frac{T i_s^2}{i_0^2(\nu)}.$$ (6.28)

Thus, the signal-to-noise ratio improves proportionally to the correlation time. It should be noted that T is half the correlation time.

6.2.2 Periodic Signal Recovering by Autocorrelation

We consider a periodic current signal with an unknown frequency spectrum buried in noise. This signal can be described by a series of harmonic functions like

$$f_S = \sum_s f_s(t) = \sum_s i_s \cos(\omega_s t + \varphi_s).$$ (6.29)

Performing an autocorrelation of the observed signal and its noise $f_N(t)$ we have according to (6.1)

$$\phi_{AA}(\tau) = \frac{1}{2T} \int_{-T}^{T} \left[\sum_s f_s(t) + f_N(t) \right] \left[\sum_s f_s(t+\tau) + f_N(t+\tau) \right] dt$$

(6.30)

or

$$\phi_{AA}(\tau) = \phi_{SS}(\tau) + 2\phi_{SN}(\tau) + \phi_{NN}(\tau),$$ (6.31)

where

$$\phi_{SS}(\tau) = \frac{1}{2T} \int_{-T}^{T} \left[\sum_s f_s(t) \right] \left[\sum_s f_s(t+\tau) \right] dt,$$ (6.32)

$$\phi_{SN}(\tau) = \frac{1}{2T} \int_{-T}^{T} \left[\sum_s f_s(t) \right] \left[f_N(t+\tau) \right] dt,$$ (6.33)

$$\phi_{NN}(\tau) = \frac{1}{2T} \int_{-T}^{T} \left[f_N(t) \right] \left[f_N(t+\tau) \right] dt.$$ (6.34)

The Fourier transform of signal $\sum_s f_s(t)$ becomes

$$F_S(\omega) = \sum_s \int_{-T}^{T} i_s \cos(\omega_s t + \varphi_s) e^{j\omega t} dt$$ (6.35)

or

$$F_S(\omega) = \sum_s i_s \left[\begin{array}{l} \{\dfrac{\sin(\omega_s + \omega)T}{\omega_s + \omega} + \dfrac{\sin(\omega_s - \omega)T}{\omega_s - \omega}\} \cos\varphi_s + \\[2mm] j\{\dfrac{\sin(\omega_s + \omega)T}{\omega_s + \omega} - \dfrac{\sin(\omega_s - \omega)T}{\omega_s - \omega}\} \sin\varphi_s \end{array} \right]. \tag{6.36}$$

Assuming a sufficiently large value of T with $T \gg \frac{1}{\omega_s}$ we may write (see footnote of (6.17))

$$F_S(\omega) = \sum_s \pi i_s \left[\begin{array}{l} \{\delta(\omega_s + \omega) + \delta(\omega_s - \omega)\} \cos\varphi_s + \\[2mm] j\{\delta(\omega_s + \omega) - \delta(\omega_s - \omega)\} \sin\varphi_s \end{array} \right]. \tag{6.37}$$

Substituting (6.37) into (6.2) we note that the integration of a product of two different δ-functions vanishes. Further, because the integration involves only positive values of ω the components with $\delta(\omega_s + \omega)$ also vanish. We then find for $\phi_{SS}(\tau)$

$$\phi_{SS}(\tau) = \frac{1}{2\pi T} \int_0^\infty \sum_s \pi^2 i_s^2 \delta^2(\omega_s - \omega) \cos\omega\tau \, d\omega. \tag{6.38}$$

If we now substitute $\pi\delta(\omega_s - \omega)/T = \sin(\omega_s - \omega)T/(\omega_s - \omega)T$ we get

$$\phi_{SS}(\tau) = \frac{1}{2} \sum_s i_s^2 \cos\omega_s\tau. \tag{6.39}$$

Although the final result with sufficiently long correlation does not depend on the type of random noise fluctuations, we shall for simplicity consider white noise as given by (6.15). For the white noise the probabilities associated with the current fluctuations are invariant under a shift of time. Then the cross correlation $\phi_{SN}(\tau)$ gives with the (6.15) and (6.37) according to (6.9)

$$\phi_{SN} = \left[\frac{\pi \overline{i_0^2(\omega)}}{2T} \right]^{1/2} \sum_s i_s \cos\varphi_s \tag{6.40}$$

or in terms of Hertz frequencies

$$\phi_{SN} = \left[\frac{\overline{i_0^2(\nu)}}{4T} \right]^{1/2} \sum_s i_s \cos\varphi_s. \tag{6.41}$$

For the autocorrelation of the noise we get according to (6.15) and (6.2)

$$\phi_{NN}(\tau) = \int_0^{\pi/2T} \overline{i_0^2(\omega)} \cos\omega\tau \, d\omega = \frac{\overline{i_0^2(\omega)}}{\tau} \sin\left(\frac{\pi\tau}{2T}\right) \tag{6.42}$$

or in terms of Hertz frequencies

$$\phi_{\mathrm{NN}}(\tau) = \frac{\overline{i_0^2}(\nu)}{2\pi\tau} \sin\left(\frac{\pi\tau}{2T}\right).$$ (6.43)

Finally by substituting the (6.39), (6.41) and (6.43) into (6.31) we find

$$\phi_{\mathrm{AA}}(\tau) = \frac{1}{2}\sum_s i_s^2 \cos\omega_s\tau + \left[\frac{\overline{i_0^2}(\nu)}{T}\right]^{1/2}\sum_s i_s \cos\varphi_s + \frac{\overline{i_0^2}(\nu)}{2\pi\tau}\sin(\frac{\pi\tau}{2T}).$$ (6.44)

It is seen that the noise described by the second and third term of (6.44) decreases with the correlation time T. Usually $T \gg \tau$ so that the third term can be written as $\overline{i_0^2}(\nu)/4T$. Thus at sufficiently long correlation the auto-correlation becomes free of the noise that accompanies the signal so that we end with

$$\phi_{\mathrm{AA}} = \frac{1}{2}\sum_s i_s^2 \cos\omega_s\tau.$$ (6.45)

We now calculate the Fourier transform of ϕ_{AA} with the aid of (6.6) and obtain the spectral power density

$$S_{\mathrm{AA}} = \frac{1}{2\pi}\sum_s i_s^2 \left[\frac{\sin(\omega_s + \omega)\tau_{\mathrm{m}}}{\omega_s + \omega} + \frac{\sin(\omega_s - \omega)\tau_{\mathrm{m}}}{\omega_s - \omega}\right],$$

where τ_{m} is the maximum performed delay time. It is seen that the spectral power density distribution has strong maxima for all signal frequencies ω_s and the larger τ_{m} the more they are pronounced.

To obtain the signal power, we integrate S_{AA} over the entire frequency range $[0, \omega_{\mathrm{max}}]$ and obtain

$$P = \frac{1}{2\pi}\sum_s i_s^2 \int_0^{\omega_{\mathrm{max}}} \left[\frac{\sin(\omega_s + \omega)\tau_{\mathrm{m}}}{\omega_s + \omega} + \frac{\sin(\omega_s - \omega)\tau_{\mathrm{m}}}{\omega_s - \omega}\right]d\omega.$$ (6.46)

For sufficiently large value of $\tau_{\mathrm{m}} \gg 1/\omega_s$ we may approximate $\sin(\omega_s - \omega)\tau_{\mathrm{m}}/(\omega_s - \omega)$ by $\pi\delta(\omega_s - \omega)$ and find as expected

$$P = \frac{1}{2}\sum_s i_s^2.$$ (6.47)

6.2.3 Autocorrelation of White Noise

Consider the autocorrelation of noise with a narrow band filtered out of a white noise spectrum between the frequencies f_1 and f_2. The spectral power density $S_{\mathrm{AA}} = \overline{i_0^2}(\omega)$ is constant. In the case of shot noise, for instance, we have according to (1.27) $\overline{i_0^2}(\omega) = ei_s/\pi$ Using (6.5) we get for the auto-correlation

$$\phi_{AA}(\tau) = \int_{2\pi f_1}^{2\pi f_2} \overline{i_0^2}(\omega) \cos \omega\tau \, d\omega = \frac{2\overline{i_0^2}(\omega)}{\tau} \cos 2\pi f_0\tau \sin \pi B\tau , \qquad (6.48)$$

where $f_0 = \frac{1}{2}(f_1 + f_2)$ and $B = f_2 - f_1$. In terms of Hertz frequencies we write

$$\phi_{AA}(\tau) = \overline{i_0^2}(\nu) B \cos 2\pi f_0\tau \left(\frac{\sin \pi B\tau}{\pi B\tau}\right) . \qquad (6.49)$$

If $f_{1,2} \gg B$ the autocorrelation can be regarded as a oscillating function with amplitude $\overline{i_0^2}(\nu)B \left(\frac{\sin \pi B\tau}{\pi B\tau}\right)$ which is $\overline{i_0^2}(\nu)B$ for $\tau = 0$ and depending on B the autocorrelation decreases fast with τ.

Next, we consider the case that the white noise is filtered by a sharply tuned circuit with center frequency f_0, and half-power transmission $|f_0 - f| = |\omega_0 - \omega|/2\pi = B/2$. The transmitted spectral power density is then given by

$$S_{AA}(\omega) = \overline{i_0^2}(\omega) \left[1 + \frac{4(f_0 - f)^2}{B^2}\right]^{-1} . \qquad (6.50)$$

We obtain for ϕ_{AA} by means of (6.5)

$$\phi_{AA}(\tau) = \int_0^\infty \overline{i_0^2}(\omega) \left[1 + \frac{4(f_0 - f)^2}{B^2}\right]^{-1} \cos \omega\tau \, d\omega . \qquad (6.51)$$

We substitute as new variable $z = (\omega - \omega_0)/\pi B$. The lower limit of the integration becomes $-\omega_0/\pi B$ which my be extended to $-\infty$ because the system is sharply tuned at $z = 0$. Further, on performing the integration we expand $\cos \omega\tau = \cos[(\omega - \omega_0)\tau + \omega_0\tau] = \cos(\omega - \omega_0)\tau \cos \omega_0\tau - \sin(\omega - \omega_0)\tau \sin \omega_0\tau$ and note that the second term with $\sin(\omega - \omega_0)\tau = \sin \pi B\tau z$ is an odd function of z and therefore vanishes by integration. We obtain

$$\phi_{AA}(\tau) = \pi B \cos(2\pi\tau f_0) \overline{i_0^2}(\omega) \int_{-\infty}^\infty \frac{\cos(\pi B\tau z)}{1 + z^2} dz \qquad (6.52)$$

or

$$\phi_{AA}(\tau) = \pi^2 B \cos(2\pi\tau f_0) \overline{i_0^2}(\omega) e^{-\pi B\tau} . \qquad (6.53)$$

It is seen from the above calculations that the noise only contributes to an autocorrelation for small values of τ and that the larger the bandwidth the faster its autocorrelation drops as a function of τ.

6.2.4 Spectral Power Density from Shot Noise Correlation

We consider a average current i_0 of charges moving with constant speed in a photodiode. The current fluctuations are described by $f_A(t)$. The transit time of the charges is τ_0. The autocorrelation of f_A is unequal zero for $\tau < \tau_0$ and zero for $\tau > \tau_0$, because of constant speed the overlap decreases according to $\left(1 - \frac{\tau}{\tau_0}\right)$. The autocorrelation for $\tau < \tau_0$ can then be written as

$$\phi_{AA}(\tau) = \frac{1}{\tau_0}\left(1 - \frac{\tau}{\tau_0}\right)\int_0^{\tau_0} f_A^2 \, dt. \tag{6.54}$$

For the current fluctuations (see (1.13) and (1.15)) we derive

$$\frac{1}{\tau_0}\int_0^{\tau_0} f_A^2 \, dt = \overline{i_n^2} = \frac{ei_0}{\tau_0}$$

and by substituting this into (6.54) we get

$$\phi_{AA}(\tau) = \frac{ei_0}{\tau_0}\left(1 - \frac{\tau}{\tau_0}\right). \tag{6.55}$$

Applying (6.6) we find

$$S_{AA} = \frac{2ei_0}{\pi\tau_0}\int_0^{\tau_0}\left(1 - \frac{\tau}{\tau_0}\right)\cos\omega\tau \, d\tau = \frac{ei_0}{\pi}\frac{\sin^2(\omega\tau_0/2)}{(\omega\tau_0/2)^2}, \tag{6.56}$$

which is in agreement with (1.31). Thus, the spectral power density can be calculated either from the individual micropulse currents as done in Chap. 1 or from the autocorrelation of current fluctuations.

6.2.5 Correlations of Linear Detector Systems

By calculating the cross correlation of the input and output of a linear system we apply the principle of superposition which is characterized by the impulse response function $h(t)$ or by the frequency response function which is the Fourier transform of $h(t)$ given by

$$H(\omega) = \int_{-\infty}^{\infty} h(t)e^{-j\omega t} \, dt \tag{6.57}$$

with $h(t) = 0$ for $t < 0$.

The output function of the linear system, $y(t)$, is then the convolution of the input function, $x(t)$, with the impulse response function or

$$y(t) = \int_{-\infty}^{\infty} x(t - \alpha)h(\alpha)d\alpha. \tag{6.58}$$

Substituting the inverse Fourier transform of $x(t - \alpha)$ into (6.58) and then changing the order of integration the result is equal to the inverse Fourier transform of $y(t)$ or

$$Y(\omega) = X(\omega)H(\omega). \tag{6.59}$$

According to (6.8) and (6.58) the cross correlation of $x\,(t)$ and $y\,(t)$ becomes

$$\phi_{xy}\,(\tau) = \lim_{T \to \infty} \frac{1}{2T} \int_{-T}^{T} \int_{-\infty}^{\infty} x\,(t+\tau)\,x\,(t-\alpha)\,h\,(\alpha)\,\mathrm{d}\alpha\,\mathrm{d}t \tag{6.60}$$

or

$$\phi_{xy}\,(\tau) = \int_{-\infty}^{\infty} \phi_{xx}\,(\tau + \alpha)\,h\,(\alpha)\,\mathrm{d}\alpha \tag{6.61}$$

which is the convolution of the autocorrelation function of the input. Taking the Fourier transforms on both sides we obtain

$$S_{xy}(\omega) = S_{xx}(\omega)H^*(\omega) \tag{6.62}$$

or

$$S_{yx}(\omega) = S_{xx}(\omega)H(\omega)\,. \tag{6.63}$$

Similarly the spectral density of the output gives by multiplying both sides of (6.59) by $Y^*\,(\omega)$

$$S_{yy}(\omega) = S_{xy}(\omega)H(\omega) \tag{6.64}$$

or by combining with (6.62) we get

$$S_{yy}(\omega) = S_{xx}(\omega)\,|H(\omega)|^2\,. \tag{6.65}$$

Example

Observing a noise spectrum the result depends also on the finite bandwidth of the used detector. If frequencies of the input noise power lie outside the pass band, the detector acts as a filter. Linear detectors, i.e., the response depending on frequency is proportional to the input power, as discussed in the previous Chaps. 3–5 have a frequency response function relative to their value at zero frequency that can be written in the form

$$H\,(\omega) = \frac{1}{1 + \mathrm{j}\omega\tau_\mathrm{d}}\,, \tag{6.66}$$

where τ_d is the time constant of the detector, determined by capacitance and conductivity.

Calculating the autocorrelation of the output detector noise as the result of the input white noise with spectral power density $\overline{i_0^2}\,(\omega)$ we find according to (6.5) and by using (6.65) and (6.66)

$$\phi_{yy}\,(\tau) = \int_0^\infty \frac{\overline{i_0^2}\,(\omega)\cos\omega\tau}{1 + (\omega\tau)^2}\,\mathrm{d}\omega = \frac{\pi\overline{i_0^2}\,(\omega)}{2\tau_\mathrm{d}}\mathrm{e}^{-\tau/\tau_\mathrm{d}}\,. \tag{6.67}$$

Thus the correlation time for which $\phi_{yy}\,(\tau)$ drops by $1/e$ of its initial value is τ_d. The faster the detector the shorter its autocorrelation time.

7

Signal Processing

Most applications require the amplification of the detected signal by external electronics. Standard electronics are, in general, not suitable. Dealing with weak signals the amplifier electronics must be well-designed to obtain high performance of the signal processing. But first of all it is very favorable to optimize the detector signal to start with. As we have shown for the various detectors the observed radiation is converted into a current source. The challenge is to observe linearity with the radiation input power and to generate as much current as possible in order to obtain the strongest signal to start with. Most detectors we treated so far have a load resistance in series with the detector element. We have shown that the maximum observable signal is obtained for a load resistance much larger than that of the detector element. In that case all signal current passes through the detector element and its value is measured by the voltage change over the element. Although this method is simple its disadvantage is the relatively high voltage drop over the load in order to provide the constant bias current through photon detectors and bolometers. Further, if the time constant of the detector, given by its RC-value, limits the frequency response a load resistance or shunt resistance smaller than that of the detector has to be installed with the consequence that a great deal of the available signal current is lost. Moreover in that case the linearity of the detector becomes questionable. Fortunately, these problems can be solved.

The first subject of this chapter discusses external electronics that allows the full detection of the generated signal current, avoids the high bias voltage, improves the frequency range, and amplifies the signal power. Further, the internal amplifier noise can be carefully controlled to minimize the deterioration of the signal-to-noise ratio. This can be accomplished with an operational amplifier, an all-solid-state integrated circuit. The amplified signal of the operational amplifier will then be above the level of any noise added in subsequent processing. The attractive property of the operational amplifier is its ability to amplify dc and ac signals simultaneously without phase shifts. It has a high voltage gain even above 10^5 and a high input impedance.

Having obtained an output signal at a substantial power level it may still happen that the signal-to-noise ratio, mainly determined at the output of the detector element, is too small or even much smaller than one so that any further processing with conventional electronics is useless. The noise that accompanies the signal may include white noise like shot noise and other disturbing pick-up signals that interfere with the signal of interest. It may even happen that next to the random noise of a wide frequency range coherent signals are present such as harmonics of 50 or 60 Hz power networks, broadcast signals, strong flicker noise with low frequency spikes etc.

If the signal of interest is repetitive the accompanying noise can be eliminated in a correlation process as was discussed in Chap. 6. Based on the correlation principle several advanced signal recovering instruments have been developed which resulted in detection systems with the theoretical maximum S/N-value. In particular, instruments based on cross correlation furnish remarkably powerful tools to improve the S/N-value. In this chapter various technical systems will be discussed.

7.1 Operational Amplifier

Consider an amplifier with an inverted output polarity, i.e., the sign of the output opposes the input as schematically shown in Fig. 7.1. The negative output terminal is connected to the positive input terminal of the amplifier via a feedback resistance R_f. The current i_f through R_f opposes the input current i_s. The voltage amplification factor or gain is $A(\omega)$ so that the relation between the input voltage V_{in} and the output voltage V_{out} is given by

$$V_{out} = -A(\omega)V_{in} . \tag{7.1}$$

The amplifier is usually considered to have a very high input resistance and a frequency dependent open-loop gain described by

$$A(\omega) = \frac{A}{1 + j\frac{\omega}{\omega_1}A} , \tag{7.2}$$

where $A \gg 1$ is a constant and ω_1 is the frequency for which the gain is one (unity gain bandwidth). The voltage gain falls above $\omega > \omega_1/A$ due to internal capacitance. Also the phase of the output voltage with respect to the

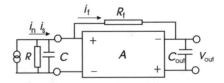

Fig. 7.1. Scheme of an inverted operational amplifier

input voltage changes above this frequency. By substituting (7.1) the output voltage becomes

$$V_{\text{out}} = \frac{-AV_{\text{in}}}{1 + j\omega\tau_{\text{a}}}, \tag{7.3}$$

where $\tau_{\text{a}} = A/\omega_1$.

The input is connected to a detector element which as we have shown can be modeled as a current generator across which there is the detector capacitance C and its internal resistance R or in parallel also a shunt resistance as shown in Fig. 7.1. The generated signal current is i_{s}. The input impedance of the amplifier is assumed much larger than that of the connected detector, whereas the output impedance is much smaller. For a FET input the impedance may be as high as 10^{12} Ω. The impedance of the detector is

$$Z_{\text{d}} = \frac{R}{1 + j\omega\tau_{\text{d}}}, \tag{7.4}$$

where $\tau_{\text{d}} = RC$. The input voltage of the amplifier is

$$V_{\text{in}} = (i_{\text{s}} - i_{\text{f}})\, Z_{\text{d}}, \tag{7.5}$$

where i_{f} is given by

$$i_{\text{f}} = \frac{V_{\text{in}} - V_{\text{out}}}{R_{\text{f}}}. \tag{7.6}$$

Solving the (7.3), (7.5), and (7.6) by eliminating V_{in} and V_{out} we find for i_{f}

$$i_{\text{f}} = \frac{Z_{\text{d}}\left(1 + \frac{A}{1+j\omega\tau_{\text{a}}}\right)}{Z_{\text{d}}\left(1 + \frac{A}{1+j\omega\tau_{\text{a}}}\right) + R_{\text{f}}}\, i_{\text{s}} \tag{7.7}$$

Using (7.4) and assuming $A \gg 1 + \omega^2\tau_{\text{a}}^2$ the real part of $Z_{\text{d}}(1 + A/(1 + j\omega\tau_{\text{a}}))$ is equal to

$$\frac{RA\left(1 - \omega^2\tau_{\text{a}}\tau_{\text{d}}\right)}{\left(1 + \omega^2\tau_{\text{d}}^2\right)\left(1 + \omega^2\tau_{\text{a}}^2\right)}.$$

By choosing the frequency and the amplification A such that

$$\frac{RA\left(1 - \omega^2\tau_{\text{a}}\tau_{\text{d}}\right)}{\left(1 + \omega^2\tau_{\text{d}}^2\right)\left(1 + \omega^2\tau_{\text{a}}^2\right)} \gg R_{\text{f}} \tag{7.8}$$

which implies $\omega^2\tau_{\text{a}}\tau_{\text{d}} < 1$ we get $i_{\text{f}} \approx i_{\text{s}}$ so that $V_{\text{in}} \approx 0$ and

$$V_{\text{out}} \approx -i_{\text{s}}R_{\text{f}}. \tag{7.9}$$

Operating the amplifier in the high gain bandwidth we have $\omega\tau_{\text{a}} \leq 1$. Therefore, the detector must fulfil the condition $\omega\tau_{\text{d}} < 1$ so that in case a shunt resistance is added to reach a higher frequency response its value decreases with increasing frequency. Then the Johnson noise current, mainly determined by the shunt resistance, increases with frequency.

Typical values for A will be in the order of 10^5 so that the above inequality can be easily satisfied in practice. From the above analysis we draw several conclusions. Firstly, the full signal current is available and the signal voltage is amplified by the factor R_f/R_d and remains linear with the signal input radiation power.

Secondly, since the output impedance of the amplifier is much smaller than that of the detector element, the signal voltage $i_s R_f$ may deliver sufficient power for further processing.

Thirdly, as long as the condition quoted with expression (7.8) remains fulfilled the voltage gain is independent on frequency and the output linear with any input pulse form.

Fourthly, the value of the amplification factor A, its frequency dependence and stability, are not relevant as long as the condition given by expression (7.8) is fulfilled.

The above described amplifier with negative feedback that drives itself to practical zero input voltage is called an operational amplifier. High input impedance and low noise are achieved by using a FET, field effect transistor, or a MOSFET, based on metal-oxide semiconductor, as the first stage of amplification. The input signal is then connected to the gate of the FET or MOSFET. With both types a high degree of control over the current flow through the transistor can be established by the electric field of the signal voltage connected to the gate. After the first stage of amplification the signal is passed on to additional stages with much lower input impedances to maintain the high frequency response. The excess noise of the amplifier, described by its noise factor F, is taken into account by the effective temperature (see Sect. 2.6). The signal voltage gain R_f/R_d of the operational amplifier is usually in the order of 10–100, a value that is in general sufficient to get above the level of any noise added in subsequent processing.

As an example, in the circuit of Fig. 7.2 a diode detector is connected in the reverse biased mode to an operational amplifier [29]. Since the input signal voltage is practically zero the measured output voltage, according to (7.9) equal to $i_s R_f$, is linear with the input power P_s. The position on the current–voltage characteristic with the minimum noise current can be simply

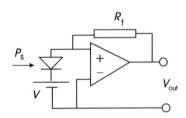

Fig. 7.2. Reverse biased diode detector connected to an operational amplifier

chosen by changing the reverse biased voltage V indicated in the figure. The best performance for small signal power is often found for the current mode operation with $V = 0$. Dealing with oscillating input signals the dc-currents like the dark current can be simply eliminated by placing a blocking capacitor in the output current circuit.

7.2 Lock-in Amplifier

The lock-in amplifier (sometimes referred to as synchronous amplifier or phase-sensitive amplifier) is an instrument that performs a cross correlation between a signal of interest and a reference signal, both at the same frequency. Any interfering noise that does not appear at the reference frequency can be averaged toward zero with sufficient long correlation time.

For many applications involving radiation detection it is straightforward to operate on the input signal with a chopper or modulator. The detected signal will then be observed at the modulating frequency. This detected signal plus a reference signal obtained from the chopper or modulator are both introduced in the lock-in amplifier, which after appropriate signal processing, provides a dc voltage output signal that is proportional to the incident beam intensity before modulation.

Let us assume that a slowly varying beam P_s is modulated and can be described by

$$P = P_s m(t),\tag{7.10}$$

where $m(t)$ describes the beam modulation with the frequency f_m. Without the applied modulation we have $P = P_s$. It is assumed that the modulation frequency f_m is much larger than any modulation frequency of P_s. The current signal delivered by the detector is then

$$i = i_s m(t),\tag{7.11}$$

where i_s is the signal produced by the beam power P_s. For example, in the case of a photodetector we have with reference to (2.3) $i_s = \eta e P_s / h\nu_s$.

The detected current signal passes first a low-noise operational amplifier. According to (7.9) the output voltage signal is equal to $i_s R_f$. The gain of this stage is usually in the order of 10–100, a value that is sufficient to get above the level of any noise added in subsequent processing. The signal is next passed through a relative narrow-band amplifier G_a that is tuned at the modulation frequency f_m. This selective amplifier eliminates already to a large degree the accompanying noise of the signal and the higher harmonics of the modulated beam. In fact it increases the dynamic range of the system by blocking the undesired signals and passing only those frequencies that are close to f_m. The bandwidth of G_a is sufficient to pass the highest significant modulation frequency of P_s with acceptable attenuation. Usually this bandwidth is not smaller than 1% of the modulation frequency f_m, otherwise it may happen that phase shifts of the relatively slow modulation frequencies of P_s will lead to an amplified signal with a time dependence that deviates from that of the incident radiation beam. If the beam is modulated with a chopper we consider rectangular pulses with amplitude P_s and a square wave $m(t)$. Expanding the square wave $m(t)$ in terms of a Fourier series we have

$$m(t) = \frac{1}{2} + \frac{2}{\pi} \left(\cos 2\pi f_m t - \frac{1}{3} \cos 6\pi f_m t + \frac{1}{5} \cos 10\pi f_m t \cdots \right). \qquad (7.12)$$

The average output signal of G_a, referring to the factor $1/2$ in (7.12), giving an output of $1/2 i_s R_f G_a$ is eliminated by the blocking capacitor.

The output signal voltage $V_g(t)$ of G_a is then obtained by taking only the first harmonic of (7.12) and we find

$$V_g(t) = \frac{2}{\pi} i_s R_f G_a \cos 2\pi f_m t. \qquad (7.13)$$

The essentials of the lock-in amplifier are shown in Fig. 7.3. The output signal of the first amplifier is followed by a double-pole reversing switch driven by an actuator which is synchronized with the modulation frequency f_m in such a way that for every polarity reversal of the signal wave there is one of the switch. In principle this polarity switch can be pure mechanical as it was at its first appearance. Modern versions apply, of course, advanced electronic components. The switch is assumed to be instantaneous.

The effect of switching can also be seen as the multiplication of the signal waveform $V_g(t)$ with a perfect square wave reference signal of zero average value and amplitude one. The two waves are shown in Fig. 7.4. The phase difference φ between the zero crossings of the two waves can be varied from 0 to 360°. If we next integrate the product of the two waves over one full period we obtain an average voltage

$$V_{av} = \left(\frac{2}{\pi} \right)^2 i_s R_f G_a \cos \varphi. \qquad (7.14)$$

Next the signal V_{av} is amplified by the second buffer amplifier with gain G_b and then lead through the resistor R to the capacitance C. After charging C the output voltage becomes

$$V_{out} = \left(\frac{2}{\pi} \right)^2 i_s R_f G_a G_b \cos \varphi. \qquad (7.15)$$

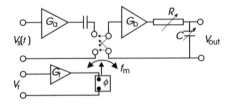

Fig. 7.3. Principle of lock-in amplifier. The amplified alternating input signal is rectified by a polarity switch. A dc output signal is obtained with a noise bandwidth determined by the low-pass RC output filter

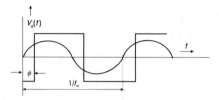

Fig. 7.4. Polarity switching of a signal is identical to the multiplication with a square wave

The result shows that the final voltage signal follows the detector signal current i_s produced by P_s. When the two signals are in phase ($\varphi = 0$) a maximum output signal is obtained. In practice the phase shifter can be accurately adjusted by hand or automatically by an electronic circuit to find the maximum signal[1].

The above described process of multiplying the signal function with the switch or reference function and then integrating the result with a capacitance is a cross correlation process as described in chap. 6.

In case the beam power P_s varies slowly compared to the chopper frequency f_m the amplitude modulation of the chopped beam is then demodulated by the cross correlation with reference signal with frequency f_m. The signal voltage of the capacitor follows the variations of P_s. The bandwidth of the low-pass RC-filter must then, of course, be sufficiently wide to respond to this amplitude modulation. This gives an upper limit to the RC-value. The lock-in amplification can therefore also be considered as the demodulation of an amplitude modulated signal beam on which a much higher single frequency modulation is superimposed in order to perform this demodulation. The charging time of C is in fact the correlation time and is set by its RC-value. This correlation time is taken as large as possible to eliminate a maximum of noise.

The detector signal and its accompanying noise, amplified by the operational amplifier, are usually much larger than the additional noise of the lock-in amplifier so that the output noise of the lock-in amplifier originates mainly from the detector. In practice, it is often possible to select the chopper frequency to a large extend at will. The choice will then be in a region that is more or less free of disturbing elements like coherent signals of the power network or flicker noise.

Let us assume that only white noise with its constant spectral power density (for instance Johnson noise and shot noise) is present and was mixed with the signal at the detector. Other noise contributions from amplifiers are negligible with respect to the amplified detector noise. The noise bandwidth

---

[1] Lock-in amplifiers may have a button which can be depressed to add 90° to the phase of the reference signal. It is experimentally easier to vary the phase shifter until a null signal is obtained. Then the 90° button is depressed to obtain the maximum in-phase signal.

in the final result is determined by the low-pass RC-filter. Its noise equivalent bandwidth for the noise power is given by

$$\Delta f_{RC} = \int_0^\infty \frac{df}{1 + (2\pi f RC)^2} = \frac{1}{4RC} . \tag{7.16}$$

As the signal with noise passes the narrow-band amplifier G_a, tuned at the frequency f_m, the amplified noise has a narrow band around f_m. The polarity switch, as we discussed, is equivalent to the multiplication with a square wave reference signal. Because the noise band of interest is set by the low-pass RC-filter we only consider the multiplication of the noise current i_n by the first harmonic of the square wave reference signal i.e., $(4/\pi)\cos 2\pi f_m t$. The spectrum of the passed noise current $(4/\pi)i_n(f)\cos 2\pi f_m t$ is split-up into a part containing sum-frequencies and another part with the difference-frequencies of which a small part near $f = 0$ will charge the low-pass filter. Only absolute values of the difference frequencies make sense so that the spectral power density of the noise at the low-pass filter originating from both negative and positive difference-frequencies given by $2(4/\pi)^2 i_n^2(f)\cos^2 2\pi f_m$ becomes $(4/\pi)^2 i_n^2(f)$. After amplification and substituting the bandwidth given by (7.16) the output noise becomes

$$\left(\overline{v_n^2}\right)_{\text{out}} = \frac{4}{\pi^2} \frac{i_n^2(f)}{RC} \left(R_f G_a G_b\right)^2 , \tag{7.17}$$

where $i_n^2(f)$ is the white noise spectral power density at the detector. The signal-to-noise ratio for maximum signal ($\varphi = 0$) derived from (7.15) and (7.17) becomes[2]

$$\frac{S}{N} = \frac{4}{\pi^2} RC \frac{i_s^2}{i_n^2(f)} . \tag{7.18}$$

Since the transmitted noise power is proportional to the bandwidth of the RC-filter it is seen that the larger RC the smaller the noise power and the larger the signal-to-noise ratio. However, the RC-value must be sufficiently small to have negligible attenuation and phase shift of the power modulation of P_s i.e., the system must be able to follow the relatively small variations of incident beam (before chopping) on the detector. A good criterion is to have $RC \leq 0.3/2\pi f_s$ where f_s is the highest significant frequency component of P_s. The amplitude attenuation of this component is then less than 5%.

Looking at the noise contribution given by 7.17 one might argue that the same result would have been obtained without modulating the beam by merely applying the low-pass filter to the amplified detector output. In that case the transmitted noise would in general originate from a very noisy region due to $1/f$-noise, harmonics of power supplies and other low frequency disturbances. The great advantage of the lock-in amplification is the shift to a low noise frequency area.

---

[2] Sometimes the S/N value is expressed as the ratio of the signal voltage to the square root of the average square of the noise voltage. In that case S/N is proportional to the square-root of the RC-value.

In conclusion, the lock-in amplifier is a complete signal processing system with high overall signal-to-noise gain, which may be as high as 10^{10}, and should be considered for any measurement of slow changing low power signal levels. The operating frequency f_{m} is usually adjustable from about 10 Hz to 100 kHz or from 10 kHz to a few MHz. The output becomes a clean reproduction, even of very small original signal voltages, which may have been as much as 60 dB below the noise level when observed by the detector. Signal sensitivities down to 10^{-6} V and current sensitivities down to 10^{-10} are feasible.

7.2.1 Two-Phase Lock-in Amplifier

If we use two identical lock-in amplifiers with the same signal connected to the input terminals and both reference channels for the polarity switches driven by the same synchronizing source which is phase-locked to the applied modulation frequency, except that the switching phase of one unit is 90° shifted with respect to the other one we obtain two average output signals which are, respectively,

$$V_{1,out} = \left(\frac{2}{\pi}\right)^2 i_{\mathrm{s}} R_{\mathrm{f}} G_{\mathrm{a}} G_{\mathrm{b}} \cos \varphi \qquad (7.19)$$

$$V_{2,out} = \left(\frac{2}{\pi}\right)^2 i_{\mathrm{s}} R_{\mathrm{f}} G_{\mathrm{a}} G_{\mathrm{b}} \sin \varphi . \qquad (7.20)$$

Taking the square root of the sum of the two squares of the output signals we obtain a phase-insensitive detection system with the output

$$V_{\mathrm{out}} = \left(\frac{2}{\pi}\right)^2 i_{\mathrm{s}} R_{\mathrm{f}} G_{\mathrm{a}} G_{\mathrm{b}} \qquad (7.21)$$

This detection technique will be applied for situations where the phase of the modulated signal drifts in a unpredictable way relative to the reference signal. For instance optical radar or communication systems where the phase depends on the distance between object and detector.

It can also be of interest to connect the two outputs to, respectively, X- and Y-axis of an oscilloscope or X–Y recorder whose X and Y gain factors are set equal and with the zero values of both signals positioned at the center of the monitor or chart. The radial distance of the recorded signal from the zero point is proportional to V_{out} whereas the angle between this radius vector and the X-axis is equal to φ. Thus it measures the in-phase and out-phase components and calculates the vector magnitude and its phase angle.

In practice the two systems are combined into one dual phase lock-in amplifier, using a single signal-tuned amplifier and two mixers in quadrature, each followed by its own separate buffer amplifier and low-pass filter system.

7.3 Signal Averagers

In many applications it is necessary to observe the actual waveform of a repetitive signal. We have seen that the lock-in amplifier resolves only the average of the rectified waveform and is therefore not suitable to analyze the waveform itself. If the waveform of interest has a trigger pulse available that is locked to the repetitive waveform, signal processing instrument, often known as signal averagers, can by used successfully for the waveform analysis.

These instruments sample each signal pulse, divide it into many increments to obtain resolution, and then store a voltage proportional to the level of each increment. By observing many repetitions a cross correlation is carried out. In this way a point by point average value of the input waveform is obtained with strong reduction of the noise.

The available trigger pulse initiates each time a sweep that will continue for any chosen duration. If in an application a stimulus signal is transmitted to observe the resulting reaction it may be convenient to use the same stimulus signal as the trigger pulse for the signal averager.

In case of high resolution the applied technique is a cross correlation between the waveform plus noise and a train of nearly delta functions with the same repetition rate. Thanks to the field effect transistor (FET) with its fast switching capability and extremely low leakage, thereby facilitating accurately holding times for on and off switching, instruments with high performance have been developed.

7.3.1 Pulse Train Averagers

An example of an instrument that applies the averaging or integrating technique is the so-called box-car integrator or single channel averager. We have seen that the lock-in amplifier is very useful for recovering continuous signals from noise. That instrument is based on beam modulation into a square wave of which only the fundamental wave of its Fourier transform is used. The subsequent technique of lock-in amplification is in principle also applicable to measure a train of short pulses of duration τ, separated by much longer time intervals T as shown in Fig. 7.5. If the repetition rate is constant the pulse train can also be resolved in a Fourier spectrum as we did with the square wave for the lock-in amplifier. However, the problem is that the content of the fundamental wave covers the fraction $4 \sin \pi \frac{\tau}{T} / \pi^2$ of the rectangular wave which has the maximum $4/\pi^2$ for the square wave with $\tau/T = 1/2$ as we also find in the result given by (7.15). The fraction approaches $4\tau/\pi T$ for small values of τ/T. Thus in the case of a pulse train with small values of τ/T only a small fraction of the signal current is used and by far most of it is wasted. Hence the lock-in amplifier is inefficient and not fruitful to recover a pulse train with a noisy background.

Fig. 7.5. Pulse train signal

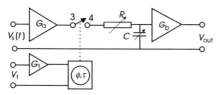

Fig. 7.6. Principle of pulse train averager or boxcar integrator. An external trigger pulse takes care of the appropriate switching between the contacts 3 and 4 in such a way that the charging of the low-pass RC-filter only occurs during the presence of the pulse. The pulse duration τ and the phase ϕ must then be adjusted

A better way to recover pulses from noise is by switching the input signal to a low-pass RC-filter on and off through the contacts three and four by means of a trigger signal V_t as indicated in Fig. 7.6. The switching is performed electronically with fast FET transistors. To close the contacts a trigger is applied to its terminals within nearly zero time to effect the transition between the contacts three and four from on to off and similarly from off to on. Because the electronic switch is an inherently noisy device it is located between the input amplifier G_a and the output buffer amplifier G_b where it will operate on signals amplified well above the intrinsic noise of the gates. The output buffer amplifier G_b is on the output side of the low-pass filter to prevent read-out circuitry from loading the capacitor. The trigger circuitry provides means by which the duration of the on-period can be chosen and its turn-on can be delayed for an adjustable time interval after the occurrence of the trigger pulse. The signal pulses themselves can also be used as trigger pulses if they are above the noise level. If the turn on and off and the delay are properly chosen by the phase φ so that each pulse of duration τ lies completely within the switch on occurring at the start of the pulse and the off exactly at the finish the charging time of the capacitance is virtually filtered. The capacitance experiences its charging by a continuous signal voltage of pulses located next to each other. Thus the boxcar integrator is a time averager of the signal over the on-intervals of the switch. Since each pulse duration is a fraction $\delta = \tau/T$ of the real repetition time the low-pass filter has an effective time constant T_{eff} equal to

$$T_{\text{eff}} = \frac{RC}{\delta} . \tag{7.22}$$

The accompanying noise of the signal pulse train is accumulated very similar as the signal. Noise spectrum components that differ in frequency with the spectrum components of the signal within the bandwidth of the low-pass filter will be mixed with the signal. Since the RC circuit is a frequency domain filter with the effective time constant T_{eff} we find with (7.16) that the effective noise equivalent bandwidth for the noise power is given by

$$\Delta f_{\text{eff}} = \frac{\delta}{4RC} . \tag{7.23}$$

Although the noise bandwidth has been effectively reduced by the gated low-pass filter the same happens to the signal bandwidth so that the ratio of the signal bandwidth to the equivalent noise bandwidth is the same as in the ungated low-pass filter. The signal-to-noise ratio is given by

$$\frac{S}{N} = \frac{4RC}{\delta} \frac{i_s^2}{i_n^2(f)} , \qquad (7.24)$$

where $i_n^2(f)$ is the spectral power density of the white noise at the detector and i_s is the signal current of a pulse.

It should be noted that unlike the lock-in amplifier the boxcar has the ability to detect also recurrent signals with irregular periods, provided that a trigger pulse precedes each occurrence by the same time advance.

7.3.2 Waveform Analyzer

The waveform of a repetitive signal $V_s(t)$ can be analyzed by slicing the wave into a large series of adjacent pulses of width τ as shown in Fig. 7.7 and then by averaging each pulse. This can be done, in principle, by a set of identical boxcars, each connected to the same signal source and activated by the same trigger train, each having the same gate width τ, such that the signal period is just equal to the gate width times the number of boxcars or $T = N_b\tau$. The delay of the gates is set in such a way that the systems turn on in a regular sequence: the first at $D = D_0$, the second at $D = D_0 + \tau \ldots$ and the nth at $D = D_0 + (n-1)\tau$. Thus on each repetition of the wave samples are taken of the N_b points. The average value of each slice of the wave will then be obtained and the accompanying noise will be filtered out with the sampling periods. The resolution of the waveform increases with the number of slices. Since the analyzer averages each of the N_b slices of the waveform the output displays a stepwise approximation of the signal wave. Depending on the type of instrument τ may be even smaller than one microsecond and N_b more than one million.

The whole analyzing technique can be performed with only one input amplifier G_a, one output buffer amplifier G_b, one trigger circuit, one series resistor R but N_b low-pass filter capacitances, each with its own gate and separate delay adjustment. This waveform analyzer or multichannel analyzer is schematically shown in Fig. 7.8.

A read-out of the waveform can be done simultaneously with the sampling or separately with all channels isolated from the input circuitry. The output

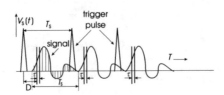

Fig. 7.7. A repetitive waveform sliced into a large series of adjacent pulses

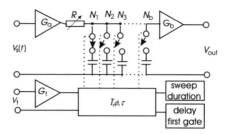

Fig. 7.8. Principle of waveform analyzer. The signal voltage of each slice of the repetitive waveform charges repetitively a channel capacitor. The waveform duration T, the pulse duration τ of a slice, and the phase ϕ of the waveform must be adjusted

signal can, for instance, be displayed while sampling so that the process of signal recovery from the background noise can be observed.

The effective time constant of each channel is RC/δ, the same as in the case of the boxcar, where $\delta = \tau/T$ is the time fraction of each channel during the wave period. The effective time constant of the whole system is also the same and the effective noise equivalent bandwidth for the noise power is therefore also given by (7.23). Thus for a given noise equivalent bandwidth the sampling time of the analyzer is independent on the number of channels.

The application of the waveform analyzer depends on the developments in the field of microelectronics. With fast switching and miniaturization to obtain many channels the technique can for instance be applied to recover photographs consisting of many pixels buried in noise that have been transmitted over long distances by electromagnetic waves (see Sect. 8.8.2). At the transmission side a waveform is obtained from the read-out of all pixels. This is done repetitively and applied as the amplitude modulation of a carrier wave. At the receiving side the beam is demodulated to get the waveform again. Then the weak repetitive noisy waveform is processed by the analyzer. Each slice of the waveform that corresponds to one pixel is stored into one channel. After sufficient sampling periods the read-out of the analyzer will show a clean picture.

7.4 Correlation Computer

Correlation computers are the most general form of signal processing equipment. They are applied to both autocorrelation and cross correlation. The above discussed lock-in amplifiers and signal averagers are in fact special cases of cross correlation equipment because a reference signal is required.

Autocorrelation analysis with a correlation computer allows the analysis of periodic signals buried in noise without the restriction of applying a synchronized reference signal. This technique can also be applied to cross correlation with the ability to describe the degree of conformity between two different signals as a function of their mutual delay.

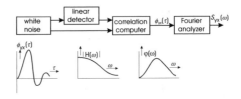

Fig. 7.9. Measuring the frequency response of an operating detector by cross correlation. The added white noise at the input of the detector is correlated with the detector output. The frequency response of the detector and its phase delay are then obtained from the Fourier transform of this cross correlation function

For any correlation the input signal is multiplied by the delayed version of itself or other signal and integrated over the correlation time as a function of the delay time between the two signals. The calculations are carried out simultaneously. The incoming signal value is multiplied by each of many discrete previous values of the same or other signal and each product is stored in the computer memories. In each memory, labeled by its specific delay time, the integration of the products is obtained over the total correlation period. A read-out of these memories can be done simultaneously with the correlation process so that the observed waveform of the correlation function as function of the delay time can be displayed. Then one observes that the noise continuously decreases and the "cleanness" of the correlation function improved with the correlation time. In case of autocorrelation and dealing with white noise the observed waveform as a function of the correlation time will be in agreement with (6.44).

Next, the correlation function on its turn can be converted into the spectral power density by calculating the Fourier transform according to (6.6) or (6.13).

In case of autocorrelation this spectral analysis is a convenient and useful means of characterizing signals with unknown frequencies. The spectral density, however, does not contain phase information of the frequency components.

With cross correlation both the sine and cosine functions must be multiplied by the correlation function at each frequency according to (6.13). The complex Fourier transform delivers the cross spectral power density of the input signals as given by (6.13) and (6.10).

Example

The frequency response of a linear system while operating in a circuit can be analyzed by adding white noise to its input. Since white noise has constant spectral power density the frequency response $H(\omega)$ is, according to (6.63), obtained by cross correlating the input white noise against the output of the system and taking the Fourier transform. This correlation result contains only the contribution from the input random white noise and not from any other source. A frequency response given by the amplitude of $H(\omega)$ and corresponding phase plot are then obtained from the Fourier analyzer as illustrated in Fig. 7.9.

8

Heterodyne Detection

For any signal processing technique there is always a fundamental level of noise below which we cannot go. The ultimate limit is known as the quantum limit of optical detection. The signal processing techniques discussed so far will not approach this limit. To reach the theoretical limit we have to suppress all noise from amplifier, dark current, and background. This can be achieved with optical heterodyne (sometimes called coherent) detection.

This technique differs significantly from direct (incoherent) detection. The principle is based on the mixing of the receiving signal with a coherent signal of a laser beam, called local oscillator, as shown schematically in Fig. 8.1. By means of a beam splitter the two beams coincide and the detector is then illuminated by the local oscillator with frequency ω_0 and a signal with frequency ω_s. Let us consider the two waves with parallel field amplitudes and propagating normal to a detector surface. As the detector measures radiation intensity which is proportional to the square of the amplitude of the total field the output signal contains a component with the difference-frequency between the monochromatic laser and the signal radiation. Thus the detector produces a signal at the difference frequency, often called heterodyne or intermediate frequency $|\omega_0 - \omega_s|$. Choosing the local oscillator much stronger than the signal the sensitivity is considerably greater than in the case of straight or incoherent detection.

Investigating broad spectrum radiation narrow-frequency bands can be selected with a narrow-band tunable amplifier. Only the radiation part that has frequency differences with the local oscillator equal to that of the pass-band of the amplifier will be detected. Thus high power sensitivity and high frequency selectivity are feasible. In addition the heterodyne receiver has antenna properties with strong directivity. Depending on the quantum efficiency of the detector the noise equivalent power becomes as low as a few times the theoretical quantum limit of $h\nu\Delta\nu$ which is a few photons per resolving time. The detector is signal limited.

However, the available spectral power density, defined as signal power per unit frequency bandwidth, must be high and at least several photons which

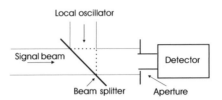

Fig. 8.1. Coinciding local oscillator and signal beam with a beam splitter

is many orders of magnitude larger than what is required for incoherent detection. This requirement is a consequence of the detection principle that the signal bandwidth is equal to that of the detector amplifier. For incoherent detection the bandwidth of the detected signal is a free choice independent of the electronic bandwidth of the detector amplifier so that the signal power can always be increased by taking a larger radiation collecting aperture or by selecting a larger bandwidth at the expense of spectral resolution.

Next to the above-mentioned fundamental requirement of high-power spectral density we have to fulfill the technical condition of high frequency stability of the local oscillator, which is usually a laser, for obtaining high spectral resolution. Unfortunately lasers are very sensitive to thermal expansion so that high frequency stability is difficult to achieve. High performance lasers have $\delta\nu/\nu$ in the range of 10^{-8}–10^{-9}. Depending on the laser frequency the fluctuations may be in the order of 10^5–10^6 Hz.

8.1 Analysis of Signal Conversion and Noise

We start with the simple case of considering the signal and local oscillator waves as plane with parallel fields to the detector surface at normal incidence. The detector efficiency is constant over its surface. The total field amplitude E_t is given by

$$E_t = E_0 \cos \omega_0 t + E_s \cos \omega_s t , \tag{8.1}$$

where the phases of the fields have been omitted because they are not relevant in this treatment. E_0 and E_s are the amplitudes of the local oscillator and signal waves, respectively. The generated current from the detector is proportional to the radiation power or to the square of the field

$$i = \beta A E_t^2 = \beta A \left[\begin{array}{l} E_0^2 \cos^2 \omega_0 t + E_s^2 \cos^2 \omega_s t + E_0 E_s \cos(\omega_0 - \omega_s) t \\ + E_0 E_s \cos(\omega_0 + \omega_s) t , \end{array} \right] , \tag{8.2}$$

where A is the size of the detector aperture and β the constant of proportionality. Since the detector cannot follow the instantaneous intensities at high frequencies it will respond to the average values of $\cos^2\omega_0 t$, $\cos^2\omega_s t$, and $\cos(\omega_0 + \omega_s) t$ which give $1/2$, $1/2$, and 0, respectively. It is assumed that the frequency response of the detector is sufficient to follow the part of the radiation power at the difference-frequency $|\omega_0 - \omega_s|$. Thus the current response of the detector is given by

$$i = \beta A \left[\frac{1}{2} E_0^2 + \frac{1}{2} E_s^2 + E_0 E_s \cos(\omega_0 - \omega_s) t \right] = i_{dc} + i_{if} . \qquad (8.3)$$

Eliminating the constant part of the current the remaining current at the heterodyne frequency, usually called the intermediate frequency, ω_{if}, is then given by

$$i_{if} = \beta A E_0 E_s \cos \omega_{if} t , \qquad (8.4)$$

where $\omega_{if} = |\omega_0 - \omega_s|$. Assuming a relatively strong local oscillator beam, $E_0^2 \gg E_s^2$, so that we may substitute $i_{dc} = i_0 = \frac{1}{2}\beta A E_0^2$, it follows that

$$i_{if} = 2 i_0 \frac{E_s}{E_0} \cos \omega_{if} t . \qquad (8.5)$$

The mean square detector current is then given by

$$\overline{i_{if}^2} = 2 i_0 i_s , \qquad (8.6)$$

where i_0 and i_s are the current generated by the local oscillator and signal beam, respectively. This result is derived for two mixed waves having the same polarization direction. In the case of arbitrary polarization of the incoming signal the beam can be considered as two parts; one with polarization parallel and the other with polarization orthogonal to the polarization of the local oscillator. Only the part with parallel polarization will have a component with the intermediate frequency. Thus if the incoming signal is randomly polarized with a uniform distribution only half of its power will be detected. The coincidence of the two beams can be obtained by means of a splitter as shown in Fig. 8.1. The detector surface is given by the size of the aperture in front of the detector.

The noise current in the detector arises in the same manner as it did in the case of straight detection discussed in previous chapters. It contains in general the amplifier noise and shot noise produced by the signal, the local oscillator, the dark current, and background radiation. For instance in the case of using a diode operating in the reverse-biased mode the noise current is given by

$$\overline{i_n^2} = 2eB \left(i_s + i_0 + i_d + i_b \right) + \frac{4k T_{eff} B}{R_L} . \qquad (8.7)$$

By taking the local power sufficiently large its noise current equal to $2eBi_0$ becomes much larger than the sum of the other noise currents so that the noise can be written as

$$\overline{i_n^2} = 2eBi_0 . \qquad (8.8)$$

Hence, the signal-to-noise ratio obtained from (8.6) and (8.8) becomes

$$\frac{S}{N} = \frac{\overline{i_{if}^2}}{\overline{i_n^2}} = \frac{i_s}{eB} = \frac{\eta P_s}{h\nu B} . \qquad (8.9)$$

Thus the minimum detectable power or NEP for S/N=1 gives

$$\mathrm{NEP}_{\mathrm{coh}} = \frac{h\nu B}{\eta}, \qquad (8.10)$$

which is equal to the theoretical limit set by the photon fluctuations, because the detector cannot observe less than $1/\eta$ photons per resolving time. The chosen bandwidth B also determines the frequency resolution, of the measurement. The great advantage of the heterodyne technique is its high frequency resolution, which is usually many orders of magnitude larger than in the case of incoherent detection.

It is seen in the above derivations that both signal and noise are proportional to the local oscillator power which drops in the S/N ratio. It should be noted , however, that the desired local oscillator power is limited by the damage threshold of the device. For instance, pyroelectric devices are not suitable as we will discuss further on.

Another fundamental restriction of coherent detection is its narrow spectral range that can be covered. This is limited either by the response time (frequency range) of the detector element or by the tunability of the laser. However, for the spectral area covered by stable and tunable lasers coherent detection surpasses in many aspects the pure optical methods like interferometry. Its resolving power and sensitivity are many orders of magnitude greater provided that at least several photons per resolving time are available. However, the latter condition is especially for incoherent radiation difficult to fulfill.

Signals from incoherent sources will in general not meet this power requirement. Radiation gathering instruments like a telescope will not help because as we discuss in Sect. 8.2 the effective aperture of the heterodyne system is limited to a single spatial mode.

8.2 Signal Beam Profile

Next to the frequency and amplitude stability of the system the alignment requires special attention. The appearance of the desired power modulation of the mixed beams at the difference-frequency of signal and local oscillator is very sensitive to the wavefront alignment of these two beams. This means a strong directivity of the input signal with respect to the local oscillator. In case of an arbitrary incident beam profile the system is selective by receiving only waves that are nearly parallel to that of the local oscillator. In fact the detection system operates also as an antenna. The first question to investigate is the dependence of the heterodyne signal on the angle between the two wavefronts and its relation with the receiving aperture of the device. To quantify this directional sensitivity it is custom to express its effect in terms of an effective detector aperture A_{eff} for a plane signal wave whose arrival direction makes an angle with that of the local oscillator.

In Sect. 8.1 we have considered for the sake of simplicity the special condition of parallel amplitudes of plane waves with normal incidence on the detector surface. Let us now consider a local oscillator wave with spatial variations. We use a reference plane with $z = 0$ near and parallel to the detector surface as indicated in Fig. 8.2. The main direction of the local oscillator wave is along the z-direction. The beam has even symmetry and the maximum intensity in the center $(x = 0, y = 0)$. The y-axis is in the plane of the drawing and the x-axis is perpendicular to it. The local oscillator wave at the reference plane has spatial variations in amplitude and phase described by $E_0(x, y)$. An incident plane signal wave with amplitude E_s has the same polarization direction as the local oscillator and propagates at first also along the z-direction. Looking at (8.4) we now have to integrate over the reference plane to obtain the complex heterodyne signal

$$i_{if} = \beta \iint E_0(x, y) E_s \cos \omega_{if} t \mathrm{d}x \, \mathrm{d}y = \beta \cos \omega_{if} t E_0(0) E_s A_{eff}(0) , \quad (8.11)$$

where $E_0(0)$ is the maximum local oscillator field amplitude in the center $(x = 0, y = 0)$ and $A_{eff}(0)$ is the effective surface area for parallel beams given by

$$A_{eff}(0) = \frac{\iint E_0(x, y) \, \mathrm{d}x \, \mathrm{d}y}{E_0(0)} . \quad (8.12)$$

The mean square detector current is then given by

$$\overline{|i_{if}|^2} = 2 i_s i_0(0) , \quad (8.13)$$

where

$$i_0(0) = \frac{1}{2} \beta E_0^2(0) |A_{eff}(0)| \quad (8.14)$$

and

$$i_s = \frac{1}{2} \beta E_s^2 |A_{eff}(0)| . \quad (8.15)$$

If this mixed field distribution at the reference plane is optically imaged with amplification γ, the intensities of signal and local oscillator will then both change by the factor γ^{-1} and their fields by $\gamma^{-1/2}$ because of conservation of

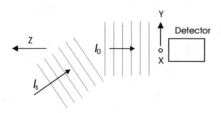

Fig. 8.2. Plane signal wave making an angle with a plane local oscillator wave

energy. Substituting the image values of E_s and E_0 and their surface areas into (8.11) we find i_{if} invariant provided all the impinging fields reach the image plane and that their optical components are lossless. Thus focusing the mixed beam on a tiny detector surface does not, in principle, deteriorate the heterodyne current signal.

Next we ask what is the contribution to the intermediate signal of a plane signal wave with the same polarization direction as the local oscillator wave and making a small angle with the direction of the local oscillator wave as indicated in Fig. 8.2. This plane signal wave with wave vector $k_s = (k_x, k_y, k_z)$ is then described by $E_s(x, y, z) = E_s e^{-jk_s r}$. At the reference plane we have

$$E_s(x, y, z = 0) = E_s e^{-j(k_x x + k_y y)} . \tag{8.16}$$

If the signal wave vector amplitude is $k_s = \omega_s/c$ and if the direction of the wave is described by circular coordinates θ and φ where $\theta = 0$ is along the z-axis and φ measures from the x-axis in the x–y plane we have

$$k_x = k_s \sin\theta\cos\varphi , \tag{8.17}$$

$$k_y = k_s \sin\theta\sin\varphi . \tag{8.18}$$

The heterodyne current is again obtained by integrating the product $E_0(x, y) E_s(x, y, z = 0)$ over the reference plane. Looking at (8.11) we now write the complex heterodyne signal current or if photocurrent as

$$i_{if} = \beta \iint E_0(x, y) E_s e^{-j(k_x x + k_y y)} \cos\omega_{if} t dx\, dy \tag{8.19}$$

$$= \beta \cos\omega_{if} t E_0(0) E_s A_{eff}(k_x, k_y) , \tag{8.20}$$

where the effective surface area $A_{eff}(k_x, k_y)$, obtained by integrating over the reference plane, is given as

$$A_{eff}(k_x, k_y) = \frac{\iint E_0(x, y) e^{-j(k_x x + k_y y)} dx\, dy}{E_0(0)} . \tag{8.21}$$

We now wish to evaluate the integrated value of $A_{eff}(k_x, k_y)$ over all possible directions of arrival of the incident plane signal wave in terms of the solid angle Ω. The differential of the solid angle is $d\Omega = \sin\theta d\theta\, d\varphi$. Using the polar coordinates in the x–y plane we have $k_r = k_s \sin\theta$ and $dk_r = k_s \cos\theta d\theta$. Changing in the (k_x, k_y)-plane to polar coordinates we substitute $dk_x dk_y = k_r dk_r d\varphi = k_s^2 \sin\theta\cos\theta d\theta d\varphi$. Substituting $d\Omega$ we get

$$dk_x dk_y = k_s^2 \cos\theta d\Omega . \tag{8.22}$$

We assume that $|A_{eff}(0)|$ is much larger than the square of the signal or local oscillator wavelength and that the radiation intensity of the local oscillator has even symmetry with its maximum for $x = 0$ and $y = 0$. In that case it turns out that $A_{\text{eff}}(k_x, k_y)$ is only appreciable for small values of θ and we may approximate $\cos\theta \approx 1$. We then use the relation $dk_x dk_y = k_s^2 d\Omega$. The integration $A_{\text{eff}}(k_x, k_y)$ over the solid angle Ω gives

$$\iint A_{\text{eff}}(k_x, k_y)\, d\Omega = \frac{1}{k_s^2} \frac{\iiiint_{-\infty}^{\infty} E_0(x, y)\, e^{-j(k_x x + k_y y)} dx\, dy\, dk_x dk_y}{E_0(0)}.$$
(8.23)

For the further evaluation of (8.23) we can take advantage of the standard two-dimensional Fourier transform relations similar to the one-dimensional relations given by (1.16) and (1.17). They are

$$E_F(k_x, k_y) = \iint_{-\infty}^{\infty} E_0(x, y)\, e^{-j(k_x x + k_y y)} dx\, dy,$$
(8.24)

$$E_0(x, y) = \frac{1}{(2\pi)^2} \iint_{-\infty}^{\infty} E_F(k_x, k_y)\, e^{j(k_x x + k_y y)} dk_x dk_y.$$
(8.25)

Applying relation (8.24) and then substituting from (8.25) $E_0(0) = \frac{1}{(2\pi)^2} \iint_{-\infty}^{\infty} E_F(k_x, k_y)\, dk_x dk_y$ we finally get the so called antenna theorem

$$\iint A_{\text{eff}}(k_x, k_y)\, d\Omega = \frac{(2\pi)^2}{k_s^2} = \lambda_s^2,$$
(8.26)

which was first postulated in [30].

In conclusion the heterodyne detector, seen as an antenna, has a receiving lobe that extends a solid angular field of view of $\Delta\Omega$ steradians with an effective aperture A_{eff} for sources inside this field of view and zero outside. In practice the relation between field of view and receiving aperture can be approximated as

$$A\Delta\Omega = \lambda_s^2,$$
(8.27)

where A is the cross-section of the local oscillator beam that falls on the detector.

With respect to the above result it is instructive to consider a simple example of a detector area A uniformly illuminated by a plane wave local oscillator beam. If the direction of the signal wave makes a small angle with that of the local oscillator the phase variation across the aperture should not exceed one wavelength, otherwise destructive signal current interferences occur. Thus the wavefront must not be tilted from parallelism by more than $\Delta\theta \approx \lambda/d$ where d is the diameter of the aperture. The incoming beam must be confined within the solid angle $\Delta\Omega \approx (\Delta\theta)^2 \approx \lambda^2/A$ so that we have the condition $A\Delta\Omega \approx \lambda^2$ in agreement with the theorem.

8.3 Optical System

Usually a heterodyne system contains optical elements to combine the beams and to focus them on a small detector area. It can be shown that the if current obtained for the integrated effective aperture of the signal may be calculated over any surface that completely intercepts the local oscillator and signal radiation. Its value is invariant to the detector position [31]. The medium may include lenses and reflectors provided that they are lossless.

Let us consider the configuration depicted in Fig. 8.3. The local oscillator beam coincides with the signal by means of a low reflective beam splitter such that the attenuation of the signal is small. The combined beams are focused on the detector area A_d. Dealing with a plane incoming signal wave its divergence after passing the aperture A is within the solid angle $\Delta\Omega \approx \lambda^2/A$. It satisfies the antenna theorem and all radiation passing the aperture contributes to the signal of interest. The spot size of the focused beam is $A_d \approx \Delta\Omega f^2$. The field of view of the focused beam seen from the detector surface is $\Delta\Omega_d \approx A/f^2$. We then find for the focused beam $\Delta\Omega_d A_d \approx \Delta\Omega A \approx \lambda^2$ so that also at the detector surface the antenna theorem still holds. The if current calculated across the focused spot on the detector is the same as calculated for the plane wave across the aperture A.

Parallelism between the detector surface and the if wave with frequency $|\omega_0 - \omega_s|$ is necessary to avoid phase variations of the if current across the illuminated detector area, otherwise the observed amplitude of the generated current, which is the integration over the detector area, will be less. The question arises what is the required flatness and parallelism of the detector for good performance. In Fig. 8.4 we show a warped detector surface and the incoming local oscillator wave and signal wave with parallel wavefronts. Starting from a

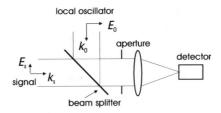

Fig. 8.3. Coinciding local oscillator and signal beam focussed on a detector

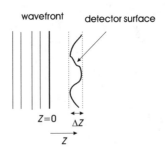

Fig. 8.4. Phase variations caused by surface roughness of the detector

wavefront at $z = 0$ where the fields are $E_0 (z = 0)$ and $E_s (z = 0)$ for the local oscillator and signal, respectively, we have after propagating the distance z the fields $E_0 (z) = E_0 (0) \, e^{jk_0 z}$ and $E_s (z) = E_s (0) \, e^{jk_s z}$. The phase of the if wave is then $(k_0 - k_s) z$. If the roughness of the surface indicated by the z-coordinate has the maximum variation Δz the phase variations over the surface will be negligible if $|(k_0 - k_s) \Delta z| \ll 2\pi$ or $|(\nu_s - \nu_0)/c| \, \Delta z = \Delta z / \lambda_{\mathrm{if}} \ll 1$.

Thus the roughness and tilt of the detector surface must be small compared with the wavelength corresponding to the if frequency. The surface needs not to be optically flat but only flat with respect to the if wavelength.

8.4 Coherent versus Incoherent Detection

8.4.1 Photodetectors

The heterodyne technique has proven to be very powerful when photodetectors are applied. Consider for instance a photodiode in the reverse-biased mode. As we have seen in Sect. 5.2.7 fast detection requires a small value for the load resistance R_L and the system is for incoherent detection amplifier limited. According to (5.84) one has

$$\mathrm{NEP}_{\mathrm{incoh}} = \frac{2h\nu_s}{e\eta} \sqrt{\frac{kT_{\mathrm{eff}} B}{R_L}}$$

in case $R_{\mathrm{sh}} \gg R_L$.

Dealing with heterodyne detection we have for the signal-to-noise ratio

$$\frac{S}{N} = \frac{2i_0 i_s}{\frac{4kT_{\mathrm{eff}} B}{R_L} + 2e i_0 B} . \tag{8.28}$$

To surpass the amplifier noise the condition for the local oscillator power is

$$P_0 \gg \frac{2kT_{\mathrm{eff}} h\nu}{e^2 R_L \eta} , \tag{8.29}$$

where we have used the relation $i_0 = e\eta P_0 / h\nu$.

Although the $\mathrm{NEP}_{\mathrm{coh}} = h\nu B / \eta$ is always much smaller than the $\mathrm{NEP}_{\mathrm{incoh}}$, the noise equivalent spectral power density of the signal for incoherent detection can be taken much smaller than in the case of heterodyne detection.

Example

For fast detection R_L is usually less than say $10^3 \, \Omega$ so that for $1 \, \mu\mathrm{m}$ radiation with $\eta = 0.73$ and $T_{\mathrm{eff}} = 600 \, \mathrm{K}$ the applied laser power should be according to (8.29) larger than $0.18 \, \mathrm{mW}$. This power requirement can be easily fulfilled and will not damage the diode. The minimum $\mathrm{NEP}_{\mathrm{coh}} = h\nu B / \eta$

becomes $2.7 \times 10^{-19}\,\mathrm{W\,Hz^{-1}}$, whereas in the case of incoherent detection we find $\mathrm{NEP_{incoh}} = 9.8 \times 10^{-12}\,\mathrm{W\,Hz^{-1/2}}$ for $R_\mathrm{L} = 10^3\,\Omega$ and $T_\mathrm{eff} = 600\,\mathrm{K}$.

Suppose the incoherent system has an optical resolution of $0.04\,\mathrm{nm}$ for $\lambda = 1\mu\mathrm{m}$ which corresponds with an optical bandwidth of 1.2×10^{10}. Choosing an electronic bandwidth $B = 10^3\,\mathrm{Hz}$ we get $\mathrm{NEP_{incoh}} = 3 \times 10^{-10}\,\mathrm{W}$. For the resolution bandwidth of 1.2×10^{10} we need a spectral power density (power per unit frequency) of the signal larger than $\frac{3 \times 10^{-10}}{1.2 \times 10^{10}} = 2.5 \times 10^{-20}\,\mathrm{J}$. This spectral power density can be further decreased by decreasing B. For the coherent case the required spectral power density of the signal must be larger than $(\mathrm{NEP_{coh}}/B) = (h\nu/\eta) = 2.7 \times 10^{-19}\,\mathrm{J}$.

8.4.2 Thermal Detector

The heterodyne technique can also successfully be applied with a thermal detector, although the frequency range is limited to values below the reciprocal thermal time constant, in practice below $100\,\mathrm{Hz}$. The generated current of the detector element is as we have seen proportional to the incident radiation power i.e. $i = \alpha P$ where α is a proportionality factor. The if signal power is given by

$$\overline{i_\mathrm{if}^2} = 2\alpha^2 P_0 P_\mathrm{s}\,, \tag{8.30}$$

where P_0 and P_s are the input powers of the laser and the signal beam. In the absence of the local oscillator we have the energy fluctuations given by $\overline{\Delta P^2} = \mathrm{NEP}^2$. The local oscillator (laser) adds thermal fluctuations described by (1.35) or

$$\overline{\Delta P_\mathrm{laser}^2} = 2h\nu P_0 B\,. \tag{8.31}$$

The current fluctuations in the thermal detector are then

$$\overline{i_\mathrm{n}^2} = \alpha^2 \left(\mathrm{NEP}^2 + 2h\nu P_0 B\right)\,. \tag{8.32}$$

For reaching the heterodyne condition we have

$$P_0 \gg \frac{\mathrm{NEP}^2}{2h\nu B}\,. \tag{8.33}$$

Let us consider the ideal thermal detector which is background limited. According to (1.84) the thermal power fluctuations are $\overline{\Delta P^2} = \mathrm{NEP}^2 = 16AB\sigma kT^5$. Substituting the background power according to (1.49) given by $P_\mathrm{B} = A\sigma T^4$ we find

$$P_0 \gg \frac{8kT}{h\nu} P_\mathrm{B}\,. \tag{8.34}$$

Taking $A = 1\,\mathrm{mm}^2$, $T = 300\,\mathrm{K}$, and a CO_2 laser with $\lambda_0 = 10.6\,\mu\mathrm{m}$ we get $P_0 \gg 7 \times 10^{-4}\,\mathrm{W}$. The minimum power P_0 will not damage the detector element. The $\mathrm{NEP_{coh}} = h\nu B = 2 \times 10^{-20}\,\mathrm{W\,Hz^{-1}}$.

Next to the thermal consideration one has to realize that the beat frequency of the local oscillator and the signal must be within the frequency range of the thermal detector. Then in view of the frequency uncertainty of lasers it will be impossible in practice to obtain a heterodyne signal.

8.4.3 Pyroelectric Detector

The situation is very different for the pyroelectric thermal detector at high frequencies. The output current at high frequencies is according to (3.63) given by $i = AK_pP/C_{th}$, where we have used the relation $\lambda\tau_{th} = C_{th}$. The amplifier noise is dominant compared with the thermal noise. Adding the shot noise of the local oscillator we have for the heterodyne case

$$\frac{S}{N} = \frac{2\left(\frac{AK_p}{C_{th}}\right)^2 P_0 P_s}{\frac{4kT_{eff}B}{R_L} + \left(\frac{AK_p}{C_{th}}\right)^2 (2h\nu BP_0)}. \tag{8.35}$$

The condition for ideal heterodyne detection is

$$P_0 \gg \frac{2kT_{eff}}{h\nu R_L}\left(\frac{C_{th}}{AK_p}\right)^2 = \frac{NEP^2}{2h\nu B}, \tag{8.36}$$

where we have used (3.72).

From the example of Sect. 3.3 we take $R_L = 7.2 \times 10^5\,\Omega$, $B = 10\,\text{kHz}$, $C_{th} = 1.64 \times 10^{-5}\,\text{J K}^{-1}$, $A = 10^{-6}\,\text{m}^2$ and $K_p = 2 \times 10^{-4}\,\text{C m}^{-2}\,\text{K}^{-1}$. We obtain $NEP^2/B = 1.44 \times 10^{-16}\,\text{W}^2\,\text{s}$. Using a CO_2 laser with $\lambda = 10.6\,\mu\text{m}$ we find $P_0 \gg 3.6\,\text{kW}$. Even for a local oscillator with $\lambda = 1\,\mu\text{m}$ the beam power must exceed $360\,\text{W}$. Thus in spite of the relatively large frequency range of pyroelectric detectors heterodyne detection is not feasible.

8.4.4 Heterodyne Detection of Incoherent Radiation

According to the antenna theorem the receiving capability of the heterodyne system is limited to $A\Delta\Omega = \lambda^2$ which is also the condition of a spatial single mode of thermal radiation reaching the detector surface A as was pointed out in Sect. 1.6. Although thermal radiation is distributed over all frequencies and all spatial modes only one single spatial mode and only frequencies within its narrow bandwidth B are detected. The radiation power of the single spatial thermal mode is $P_{th} = h\nu B/(e^{h\nu/kT} - 1)$. As found with (8.9) the minimum detectable power for $\eta = 1$ is $h\nu B$ which is for values of $e^{h\nu/kT} > 2$ more than the received thermal power. The radiation content of the receiving spatial mode will not increase if a gathering instrument like a telescope is used. The received thermal power of the heterodyne system even at high temperatures, for example $5{,}000\,\text{K}$, is much less than the minimum detectable power at optical or near infrared frequencies.

The radiation distribution of an incoherent source in a certain range of the spectrum can always be described by a temperature. The evaluation of the signal processing of incoherent radiation with coherent detection is then the same as for thermal radiation. In contrast the incoherent detection is much more favorable. The incoherent detector can receive much more power over a larger space angle and a spectral range very much broader than the narrow bandwidth B of the electronic system. Incoherent detection is therefore much more suitable for incoherent radiation. Thus the heterodyne system is insensitive for detecting incoherent radiation and so far not attractive for specific applications like astronomical spectroscopy. However, discussed in Sect. 8.5 the detection capability can be considerably increased if heterodyne detection is combined with a crosscorrelation technique. In this way high-resolution spectroscopy, limited by the electronic bandwidth, becomes feasible.

8.5 Heterodyne Lock-In Amplification

The relative high, in many cases dominating, shot noise of the heterodyne signal can be decreased by modulating or chopping the mixed optical beam and then by applying crosscorrelation. This can be accomplished with some modifications of the lock-in amplification technique for incoherent detection as discussed in Chap. 7.

We consider again the detection of incoherent radiation from an emission band with low spectral power density. The principle of lock-in amplification of the heterodyne signal is shown schematically in Fig. 8.5. Chopping the combined beams the detector output consists of a current pulse train with frequency f_m. During each optical pulse the detector will respond to the average powers of signal and local oscillator and to the mixed signal at the intermediate frequency. After passing a blocking capacitor to eliminate the constant part the heterodyne signal at the intermediate frequency is led to the operational amplifier. Its output signal $V_s(t)$, equal to the detector signal current times the output impedance R_f of the operational amplifiers, is then led to a narrow-band amplifier with gain G_a and bandwidth B tuned at the selected intermediate frequency $f_{if} = (1/2\pi)|\omega_0 - \omega_s|$. The chosen electronic bandwidth B is equal to the detected bandwidth $\Delta\nu$ of the incoherent radiation provided the frequency fluctuations of the local oscillator (laser) are much less. In contrast to incoherent lock-in amplification, discussed in Chap. 7, the signal to be processed is now a band-limited alternating signal rather than a constant signal proportional to the radiation input power. The problem for the subsequent phase-sensitive amplification is that this alternating signal exhibits during a pulse a zero average. To overcome this problem

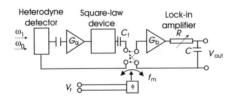

Fig. 8.5. Schematic drawing of heterodyne detector with lock-in amplifier. The radiation falling on the detector is chopped. After amplifying the heterodyne signal is converted into a positive signal by means of a square-law device. Then the chopped signal is rectified by a polarity switch and led to a low-pass RC-filter

the signal and its noise are sent to a full-wave square-law device which converts the alternating input voltage $V_{in} = G_a R_f (i_{if} + i_n)$ into a positive signal $V_{sq} = \alpha V_{in}^2$ or

$$V_{sq} = \gamma \left(i_{if}^2 + i_n^2 + 2 i_{if} i_n \right), \qquad (8.37)$$

where $\gamma = \alpha G_a^2 R_f^2$

The signal part of V_{sq} equal to $\gamma \overline{i_{if}^2}$ is a pulse train with frequency f_m of the chopper and an average (positive) voltage $\overline{V_s}$ during the pulse equal to

$$\overline{V_s} = \gamma \overline{i_{if}^2} = 2\gamma i_0 i_s, \qquad (8.38)$$

where we have substituted (8.6). Let us assume that these chopped pulses are rectangular with the intermediate time equal to the pulse duration. To obtain a square wave pulse the signal passes a blocking capacitor C_f. The obtained square wave has zero average and an amplitude equal to $0.5\overline{V_s}$. Next this wave passes a double-pole reversing switch driven by an actuator, which is synchronized with the chopping frequency f_m and in phase with the square wave by means of a reference signal V_r and an adjustable phase shifter so that the input signal of the subsequent buffer amplifier G_b is a continuous positive voltage pulse equal to $0.5\overline{V_s}$. The final signal output voltage over the capacitor C will be with (8.38)

$$V_{out} = \gamma G_b i_0 i_s. \qquad (8.39)$$

Next we consider the noise voltage described by the term $v_n = \gamma i_n^2$ of (8.37). The oscillating part of it that passes the blocking capacitor is described by $\gamma \left(i_n^2 - \overline{i_n^2} \right)$. Since this noise is filtered by the narrow bandpass we are interested in its frequency distribution. For that purpose we note that its corresponding noise power equals to $\gamma^2 \left(i_n^2 - \overline{i_n^2} \right)^2$ can also be described in terms of spectral power density.

The frequency distribution of the shot noise $\overline{i_n^2} = 2ei_0$ has the spectral bandwidth B at the intermediate frequency $f_{if} = |\nu_0 - \nu_s|$, determined by the narrow-band amplifier G_a. Let us assume that this shot noise has a constant spectral power density between the frequencies f_1 and f_2 where $f_2 - f_1 = B$ and $\frac{1}{2}(f_2 + f_1) = f_{if}$. Analyzing the subsequent noise filtering process we use the classical description of the noise spectrum.

The input noise voltage of the square-law device is equal to $v_n = i_n R_f G_a$. The square of the noise voltage at the output consists of the sum of the squares of all individual frequency components and also the sum of the products of each component with each of the other components. The latter can be split-up in components with the sum-frequencies and the difference-frequencies. The sum-frequencies run from $2f_1$ to $2f_2$ and the difference-frequencies from 0 to $f_2 - f_1$. The sum of the squares of all individual components is equal to average square of the noise voltage $\overline{v_n^2} = \overline{i_n^2} R_f^2 G_a^2$ because the average of all other components is zero. The fluctuating part of the square of the noise voltage which contains all sum- and difference-frequencies is then equal to $R_f^2 G_a^2 (i_n^2 - \overline{i_n^2})$. This implies that after the noise has passed the square-law device and the capacitor C_f we are left with a fluctuating noise voltage v_n (sq) equal to

$$v_n \text{ (sq)} = \gamma \left(i_n^2 - \overline{i_n^2} \right) . \tag{8.40}$$

The noise voltage v_n (sq) is not correlated with the reference signal V_r driving the polarity switch and is therefore fully transmitted. Thus the noise voltage being amplified by the buffer amplifier G_b becomes

$$v_{an} \text{ (sq)} = \gamma G_b \left(i_n^2 - \overline{i_n^2} \right) . \tag{8.41}$$

Taking the average square of (8.41) we get

$$\overline{v_{an}^2 \text{ (sq)}} = \gamma^2 G_b^2 \overline{\left(i_n^2 - \overline{i_n^2} \right)}^2 , \tag{8.42}$$

which is also equal to the sum of the squares of all individual field components with sum- and difference-frequencies. However, the low-filter capacitor C with its narrow band will be charged by only a small part of it. The question is then which part of the spectral density distribution of the noise will be stored.

The frequency distribution of the squares of the field amplitudes at the output of the square-law device containing the difference-frequencies is largest for $f = 0$ and decreases linearly to zero for $f = f_2 - f_1$. Similarly the frequency distribution of the squares of the field amplitudes containing the sum-frequencies has its maximum for $f_1 + f_2$ and decreases linearly on both sides of this maximum to zero for $2f_1$ and $2f_2$, respectively. In Fig. 8.6a the spectral power density is plotted that corresponds to the difference- and sum-frequencies. The integrated power under both curves of Fig. 8.6a is equal to $\overline{v_{an}^2 \text{ (sq)}}$ given by (8.42). Because the correlated amplitudes of the difference-

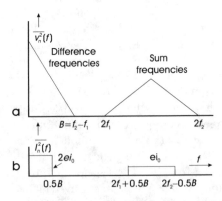

Fig. 8.6. Signal independent spectral noise power density (**a**) and signal dependent spectral noise power density (**b**) at the output of the square wave device

and sum-frequencies are equal, the integrated noise powers under the curves are also equal. The spectral density for $f = 0$ is then

$$\overline{v_n^2(0)} = \frac{1}{B}\gamma^2 G_b^2 \overline{\left(i_n^2 - \overline{i_n^2}\right)^2}. \tag{8.43}$$

Assuming that both the chopping frequency f_m and the bandpass width $1/4RC$ of the low-pass filter are much small than B we find the voltage square of the noise at the capacitor C given by

$$\overline{v_n^2} = \frac{1}{4RCB}\gamma^2 G_b^2 \overline{\left(i_n^2 - \overline{i_n^2}\right)^2}. \tag{8.44}$$

For the further evaluation of the noise we use the relation

$$\overline{\left(i_n^2 - \overline{i_n^2}\right)^2} = \overline{i_n^4} - \left(\overline{i_n^2}\right)^2. \tag{8.45}$$

The shot noise current results from the concerted action of a large number of independent producers. As pointed out in Appendix A.2.3 the statistical distribution of this noise current can therefore be described by the Gaussian probability density function

$$F(i_n) = \frac{1}{\sqrt{2\pi \overline{i_n^2}}} e^{-i_n^2/2\overline{i_n^2}}. \tag{8.46}$$

The average value i_n^4 is then obtained by calculating the integral

$$\int_{-\infty}^{\infty} i_n^4 F(i_n)\, di_n = 3\left(\overline{i_n^2}\right)^2. \tag{8.47}$$

Substituting (8.45) and (8.47) into (8.44) we get

$$\overline{v_n^2} = \frac{1}{2RCB}\gamma^2 G_b^2 \left(\overline{i_n^2}\right)^2. \tag{8.48}$$

Substituting $\overline{i_n^2} = 2ei_0 B$ we finally get

$$\overline{v_n^2} = \frac{2e^2 i_0^2 B \gamma^2 G_b^2}{RC}. \tag{8.49}$$

Next we consider from (8.37) the term $2\gamma i_{if} i_n = 2\gamma\beta AE_0 E_s i_n \cos 2\pi f_{if} t$ where we have used (8.4). The spectrum of the term $i_n(t) \cos 2\pi f_{if} t$ is split-up into a part with the frequency range between $f_{if} + f_1$ and $f_{if} + f_2$ and the part with frequencies between $f_{if} - f_1$ and $f_{if} - f_2$ where $f_{if} = 1/2\pi\omega_{if} = (1/2)(f_1 + f_2)$. Only the absolute values of the frequencies make sense so that the power spectral density of $i_n^2(t) \cos^2 \omega_{if1} t$ plotted in Fig. 8.6b becomes $2ei_0$ for $0 < f < f_2 - f_{if} = (1/2)B$, zero for $(1/2)B < f < f_1 + f_{if}$ and ei_0 for $f_1 + f_{if} < f < f_2 + f_{if}$. As the bandpass width $1/4RC$ of the lowpass filter is assumed to be much smaller than B the square of the output noise voltage for the bandwidth $1/4RC$ becomes finally

$$\overline{v_{\mathrm{n}}^2} = \frac{1}{2RC} ei_0 \left(2\gamma\beta G_{\mathrm{b}} AE_0\right)^2 E_{\mathrm{s}}^2 = \frac{8}{RC} \gamma^2 G_{\mathrm{b}}^2 ei_0^2 i_{\mathrm{s}}. \tag{8.50}$$

The signal-to-noise ratio of the low-pass output filter can now be derived from (8.39), (8.49), and (8.50). We find

$$\frac{S}{N} = \frac{i_{\mathrm{s}}^2}{\frac{2e^2B}{RC} + \frac{8ei_{\mathrm{s}}}{RC}}. \tag{8.51}$$

Looking for the NEP-value we solve (8.51) by substituting $S/N = 1$ and find

$$i_{\mathrm{s}} = \frac{4e}{RC} + \left[\left(\frac{4e}{RC}\right)^2 + \frac{2e^2B}{RC}\right]^{1/2}. \tag{8.52}$$

For $B \gg 1/RC$ and by substituting $i_{\mathrm{s}} = e\eta P_{\mathrm{s}}/h\nu$ we obtain

$$\mathrm{NEP} = \frac{h\nu}{\eta} \sqrt{\frac{2B}{RC}}. \tag{8.53}$$

Further if $i_{\mathrm{s}} \ll \frac{eB}{4}$ we obtain from (8.51)

$$\frac{S}{N} = \frac{\eta^2 P_{\mathrm{s}}^2}{2\left(h\nu\right)^2} \frac{RC}{B} \tag{8.54}$$

or in terms of signal-to-noise voltage ratio

$$\left(\frac{S}{N}\right)_V = \frac{\eta P_{\mathrm{s}}}{h\nu} \sqrt{\frac{RC}{2B}}. \tag{8.55}$$

If we now substitute $P_{\mathrm{s}} = I_{\mathrm{s}}B$ where I_{s} is the spectral power density received by the detector we get

$$\left(\frac{S}{N}\right)_V = \frac{\eta I_{\mathrm{s}}}{h\nu} \sqrt{\frac{RCB}{2}}. \tag{8.56}$$

Comparing (8.53) with (8.10) it is seen that the lock-in amplification decreases the NEP by a factor $\sqrt{2/RCB}$.

If $i_{\mathrm{s}} \gg eB/4$ or $P_{\mathrm{s}} \gg h\nu B/4\eta$ we obtain from (8.51)

$$\frac{S}{N} = \frac{P_{\mathrm{s}}}{\frac{8h\nu}{\eta RC}}, \tag{8.57}$$

which is the S/N-value of a conventional heterodyne system with bandwidth $8/RC$.

Finally we remark that if a low-power beam with a bandwidth determined by optical methods, much larger than the above-mentioned electronic bandwidth, is detected directly (incoherently) by means of photon counting combined with lock-in amplification the NEP decreases with the reciprocal value of the sampling time T as will be derived in Sect. (9.2) and given by (9.19). This decrease is much faster than in the case of coherent detection given by (8.53) because the sampling time $T = RC$ is always chosen much larger than $1/2B$. Thus if the aim is only to improve the signal-to-noise ratio and not the spectral resolution it is more attractive to apply photon counting for weak signals.

8.5.1 High-Spectral Resolution

Unfortunately the potential of high-resolution spectroscopy is at the expense of the signal-to-noise ratio. As we pointed out earlier heterodyne spectroscopy of thermal sources or more general the heterodyne detection of incoherent radiation is not simply feasible for radiation frequencies $\nu > (kT/h)\,ln2 = 1.5 \times 10^{10}\,T$. For specific applications like astronomical spectroscopy and interferometry, where high-spectral resolution is desired, an additional correlation process is needed to suppress the noise and therewith to improve the signal-to-noise ratio. With this combination high-spectral resolution can be successfully carried out. The spectral data are then obtained by accurately tuning the frequency f_{if} (equal to the optical intermediate frequency) of the narrow-band amplifier G_a. However, the detector signal at the intermediate frequency $f_{if} = |\nu_0 - \nu_s|$ contains the response of two spectral regions: one with its center frequency $\nu_0 - f_{if}$ and the other one with frequency $\nu_0 + f_{if}$ as indicated in Fig. 8.7. To eliminate one part one might think of introducing a narrow-band optical filter to block one of the two spectral regions but this is most likely very difficult to realize.

If the complex spectral region of interest with strong variations can be selected in combination with a more or less flat counter part as shown in Fig. 8.7 the signal will show the desired complexity.

Substantial additional information to disclose more details of the complexity can be obtained by also detecting the derivative of the spectral density as a

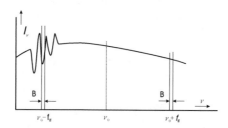

Fig. 8.7. Observed spectral regions of bandwidth B with a heterodyne system

function of frequency. This can be accomplished with a frequency-modulated local oscillator and lock-in amplification at this modulation frequency. The frequency is modulated according to

$$\nu_{sm} = \nu_0 + a \cos 2\pi f_m t, \tag{8.58}$$

where f_m is the modulation frequency and a the amplitude. Keeping the heterodyne frequency f_{if} constant the frequencies of both detected signals, below and above ν_0, are also shifted by $a \cos 2\pi f_m t$. The heterodyne signal will then contain the frequency dependent spectral density I_s integrated over the line width B or

$$P_s = I_s \left(\nu_s + a \cos 2\pi f_m t \right) B. \tag{8.59}$$

For small amplitudes a we may write

$$P_s = \left[I_s \left(\nu_s \right) + a \cos 2\pi f_m t \frac{dI_s}{d\nu} \right] B. \tag{8.60}$$

Let us now assume that the variations with frequency of the counter part are very small and can be ignored by choosing the appropriate frequency of the local oscillator. The derivative of the signal can then be found with (8.39) by sampling. However, we must realize that in (8.39) due to chopping half of the signal current was substituted. In the present case the signal at the low-pass filter is the integration of the rectified cosine-function equal to $2/\pi$. We obtain by substituting $i_s = 4(e\eta/\pi h\nu)aB(dI_s/d\nu)$ into (8.39)

$$V_{out} = \alpha G_b G_a^2 R_f^2 i_0 \left(\frac{4e\eta}{\pi h\nu} aB \frac{dI_s}{d\nu} \right). \tag{8.61}$$

By tuning the frequency f_{if} and keeping B as small as possible the derivative of the spectral power density is measured. The accompanying noise is in principle determined again only by the local oscillator and given by (8.49). We finally obtain a signal-to-noise ratio equal to

$$\left(\frac{S}{N} \right)_V = \frac{4}{\pi} \frac{\eta}{h\nu} \frac{dI_s}{d\nu} a \sqrt{\frac{RCB}{2}}. \tag{8.62}$$

In a similar way higher derivatives of I_s can be obtained by extending the expansion of (8.59). For instance the second derivative of I_s oscillates with twice the modulation frequency and can thus be solved by correlating with twice the modulation frequency in the lock-in amplifier.

8.6 Dual Signal Beam Heterodyne Detection

As pointed out in Sect. 8.4.4 heterodyne detection requires a brightness of at least several photons per unit frequency bandwidth. These powers can be delivered with laser systems. Several advanced applications based on heterodyne detection like radar, communication, and measurements of localized flow velocities in fluids use laser systems as transmitter and an optical heterodyne system with appropriate field of view and aperture as receiver. If one has the disposal of a high frequency stabilized laser the electronic detection bandwidth can be chosen very narrow so that the noise equivalent power is low. In that case high sensitivity is feasible.

Unfortunately lasers are very sensitive to thermal expansion so that high-frequency stability is difficult to achieve. High performance lasers have $\delta\nu/\nu$ in the range of 10^{-8}–10^{-9}. Depending on the laser frequency the fluctuations may be in the order of 10^5–10^6 Hz. Furthermore in several applications like radar and satellite communication the received signal is Doppler shifted by the radial velocity of the target giving uncertainty to the signal frequency. As a consequence a large electronic bandwidth of the detector must be adjusted. This effect increases with frequency because the Doppler shift is proportional to the radiation frequency. Thus the laser frequency fluctuations and the uncertain Doppler shift require broad bandwidth detection resulting in degraded sensitivity.

In this section we discuss the operation of a dual signal beam heterodyne system. This technique is a considerable improvement upon the foregoing conventional heterodyne detection because it allows narrow bandwidth detection. It was proposed by Teich [33] and experimentally studied by Abrams and White [34]. It consists of the mixing of the local oscillator with two signal beams. This three-frequency system eliminates the necessity for a highly stabilized local oscillator. It also allows targets to be continuously observed with unknown Doppler shifts while the electronic bandwidth can be taken very narrow. This is especially of interest in the short wavelength range where the Doppler shifts are large.

Since heterodyne detection requires several photons per unit frequency we consider in the following laser radiation. The required laser power depends on the distance to the target, cross-section of target, absorption, and the NEP of the system. In Fig. 8.8a block diagram of the three frequency system is shown. A laser emits two beams with frequencies ω_1 and ω_2 of which the difference-frequency $\omega_c = |\omega_1 - \omega_2|$ is accurately known. This can be obtained from a two-mode laser of which the frequencies usually fluctuate but not their frequency difference. Another way is starting from a single frequency laser which is modulated into two frequency components. The two transmitted beam may be frequency shifted by a moving target. The Doppler shifted frequency of a beam is given by

$$\omega' = \omega\left(1 \pm \frac{2v}{c}\right), \tag{8.63}$$

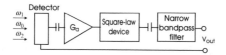

Fig. 8.8. Block diagram of a dual signal beam heterodyne detector. Two amplified heterodyne signals pass a square-law device and a capacitor with narrow bandpass filter to get a signal with the difference frequency of these two heterodyne signals

where v is the radial velocity (in the direction of the receiver) of the target and c the speed of light. The frequency difference of the two received signals is

$$\omega_c' = \omega_c \left(1 \pm \frac{2v}{c} \right) . \tag{8.64}$$

It turns out that in practice $2v\omega_c/c$ is very small and within the chosen narrow electronic bandwidth of the detection system so that we consider $\omega_c' = \omega_c$.

Another advantage of the dual system is that phase perturbations in atmospheric transmission are identical for both beams so that its degraded effect on the system performance is also considerably reduced.

Further it may happen that the target rotates so that the scatterers at one side of the target move with respect to the other side. The difference between the Doppler shifts of the two sides gives a frequency broadening to each signal. This broadening is given by $\Delta\omega_d = \omega(2D/c)\,(d\theta/dt)$ where D is the target dimension and $(d\theta/dt)$ is the angular velocity perpendicular to the beam direction. We assume in the following that this broadening effect, if present, is also within the narrow bandwidth of the detection system.

We consider that the field E_t of the incident radiation of the detector consists of three plane, parallel coincident waves with the same polarization direction. The derivation of the signal is similar to the conventional two-frequency system treated in Sect. 8.1. We get

$$E_t = E_0 \cos\left(\omega_0 t\right) + E_1 \cos\left(\omega_1 t\right) + E_2 \cos\left(\omega_2 t\right) , \tag{8.65}$$

where E_1 and E_2 are the signal fields. $E_0 \gg E_1, E_2$ is the field of the local oscillator. We have neglected the random phases of the beams because they are not relevant in this treatment. The generated detector current is

$$i = \beta A E_t^2 , \tag{8.66}$$

where A is the size of the detector aperture and β a constant of proportionality. The signal current consists of a constant part and oscillating parts that contain the terms $E_0 E_1$ and $E_0 E_2$. The term containing $E_1 E_2$ is negligible. The detector will not respond to the high-optical frequencies. The average constant values will be filtered out by a blocking capacitor. The output current at the intermediate frequencies is in analogy with (8.4) given by

$$i_{if} = i_{if1} + i_{if2} , \tag{8.67}$$

where $i_{if1} = \beta A E_0 E_1 \cos\left(\omega_{if1} t\right)$, $i_{if2} = \beta A E_0 E_2 \cos\left(\omega_{if2} t\right)$, $\omega_{if1} = \omega_0 - \omega_1$ and $\omega_{if2} = \omega_0 - \omega_2$. Assuming either $\omega_0 > \omega_1, \omega_2$ or $\omega_0 < \omega_1, \omega_2$ we have

$$\omega_c = |\omega_{if1} - \omega_{if2}| = |\omega_1 - \omega_2| . \tag{8.68}$$

With $E_0^2 \gg E_1^2, E_2^2$ so that $i_{dc} = i_0 = (1/2)\beta A E_0^2$ we obtain for the mean square of i_{if}

$$\overline{i_{if}^2} = 2i_0 \left(i_1 + i_2 \right) , \tag{8.69}$$

where $i_1 = (1/2)\beta AE_1^2$ and $i_2 = (1/2)\beta AE_2^2$. The dominating noise is again the shot noise produced by the local oscillator or

$$\overline{i_n^2} = 2ei_0\Delta f \,, \tag{8.70}$$

where Δf is the width of the bandpass filter of the amplifier G_a indicated in Fig. 8.8. At the output of the amplifier we obtain for the signal-to-noise ratio

$$\frac{S}{N} = \frac{i_1 + i_2}{e\Delta f} = \frac{\eta(P_1 + P_2)}{h\nu\Delta f} \,, \tag{8.71}$$

This result is similar to (8.9) except that now $P_1 + P_2$ is the total input signal power.

The detector signal passes in sequence the operational amplifier which gives an output signal voltage $R_f i_{if}$ and the amplifier G_a. Next the voltage signal is sent to the full-wave square-law device. The input voltage is

$$V_{in} = G_a R_f(i_{if} + i_n) \,, \tag{8.72}$$

where i_n is the noise current. The output voltage of the square-law device with $V_{out} = \alpha V_{in}^2$ becomes by substituting (8.67) and (8.72)

$$V_{out} = \gamma \left[i_{if1}^2 + i_{if2}^2 + i_n^2 + 2i_{if1}i_{if2} + 2i_n i_{if1} + 2i_n i_{if2} \right] \,, \tag{8.73}$$

where $\gamma = \alpha G_a^2 R_f^2$. To eliminate the constant part $i_{if1}^2 + i_{if2}^2$ of V_{out} a second blocking capacitor is introduced in the circuit as shown in Fig. 8.8.

The term $2i_{if1}i_{if2}$ contains the sum- and difference-frequency of ω_{if1} and ω_{if2}. The signal of interest with the frequency $f_c = \frac{1}{2\pi}|\omega_{if1} - \omega_{if2}|$ and voltage $\gamma\beta^2 A^2 E_0^2 E_1 E_2 \cos(\omega_{if1} - \omega_{if2})\,t$ will now be filtered out by tuning the narrow bandpass filter. Since f_c is known with great accuracy the bandwidth at f_c can be narrow. The average square of the output voltage at the frequency f_c is

$$V_s^2 = 8\gamma^2 i_0^2 i_1 i_2 \,. \tag{8.74}$$

Next we consider from (8.73) the noise voltage described by the term $v_n = \gamma i_n^2$. The oscillating noise part of it that passes the blocking capacitor is equal to $\gamma(i_n^2 - \overline{i_n^2})$. As this noise is filtered by the narrow bandpass we are interested in its frequency distribution. For that purpose we note that the corresponding noise power equal to $\overline{v_n^2} = \gamma^2 \overline{(i_n^2 - \overline{i_n^2})}^2$ can also be described in terms of spectral power.

Since the shot noise of the heterodyne system is Gaussian we find by applying (8.45) and (8.47)

$$\overline{\left(i_n^2 - \overline{i_n^2}\right)^2} = 2\left(\overline{i_n^2}\right)^2 \,. \tag{8.75}$$

Analyzing the spectral power density of $2\left(\overline{i_n^2}\right)^2$ we realize that the bandwidth of the shot noise was set by the bandpass filter of the amplifier G_a with

$\Delta f = f_{\mathrm{n}} - f_l$ where f_{n} and f_l are the upper and lower cutoffs, respectively. The frequency range Δf must cover the intermediate frequencies $f_{\mathrm{if1}} = \frac{1}{2\pi}\omega_{\mathrm{if1}}$ and $f_{\mathrm{if2}} = \frac{1}{2\pi}\omega_{\mathrm{if2}}$. However, these frequencies can be Doppler shifted and their range may be uncertain. Therefore it is necessary to use a large bandwidth Δf.

We now follow the same discussion as given in Sect. 8.5. The square of the noise current consists of the squares of all individual frequency components equal to $\overline{i_{\mathrm{n}}^2}$ and the sum of the products of each component with each of the other components. The latter part that passes the blocking capacitor can be written in terms of sum-frequencies and difference-frequencies. The sum-frequencies run from $2f_1$ to $2f_{\mathrm{n}}$ and the difference-frequencies from 0 to $f_{\mathrm{n}} - f_1$. The average of the sum of the squares of all frequency components is then equal to $2\left(\overline{i_{\mathrm{n}}^2}\right)^2$. The frequency distribution of the average values of the squares of the components containing the difference-frequencies is largest for $f = 0$ and decreases linearly to zero for $f = f_{\mathrm{n}} - f_1$. Similarly the distribution of the average values of the squares of the components containing the sum-frequencies has its maximum for $f = f_{\mathrm{n}} + f_1$ and decreases linearly on both sides to zero for $f = 2f_1$ and $f = 2f_{\mathrm{n}}$, respectively. This is depicted in Fig. 8.9 where we plot the spectral power density $\overline{v_{\mathrm{n}}^2}(f)$ of the noise power $\gamma^2\left(i_{\mathrm{n}}^2 - \overline{i_{\mathrm{n}}^2}\right)^2$. The triangles OAC and DEF represent the difference- and sum-frequencies, respectively. Because the correlated amplitudes of the difference- and sum-frequencies are equal the integrated noise power indicated by the area sizes of the triangles are also equal. Thus the integrated power spectral density of the difference-frequencies v_{nD}^2 indicated by the area of the triangle OAC is equal to $\gamma^2\left(\overline{i_{\mathrm{n}}^2}\right)^2$. From this we find that its spectral power density for $f = 0$ becomes

$$\overline{v_{\mathrm{n}}^2(0)} = \frac{2\gamma^2\left(\overline{i_{\mathrm{n}}^2}\right)^2}{\Delta f}. \tag{8.76}$$

The noise that passes the narrow bandwidth B at the tuned frequency f_{c} is also indicated in Fig. 8.9. Since B is much smaller than Δf the average square of the noise voltage within the bandwidth B becomes

$$\overline{v_{\mathrm{n}}^2} = \frac{\Delta f - f_{\mathrm{c}}}{\Delta f}\frac{2\left(\overline{i_{\mathrm{n}}^2}\right)^2}{\Delta f}\gamma^2 B. \tag{8.77}$$

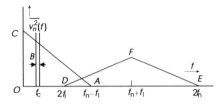

Fig. 8.9. Signal independent spectral noise power density at the output of the square wave device. The triangles OAC and DEF represent the spectral power densities that correspond to the difference- and sum-frequencies of the noise band, respectively

Substituting $\overline{i_n^2} = 2ei_0\Delta f$ we get

$$\overline{v_n^2} = 8\gamma^2 e^2 i_0^2 \left(\Delta f - f_c\right) B \,. \tag{8.78}$$

Next we consider the two last terms of (8.73) in which the noise current $i_n(t)$ is multiplied by the signals at the intermediate frequencies. The spectrum of the term $i_n(t)\cos\omega_{if1}t$ is split-up into a part with the frequency range between $f_1 + f_1$ and $f_1 + f_n$ and the part with frequencies between $f_1 - f_1$ and $f_1 - f_n$ where $f_1 = \frac{1}{2\pi}\omega_{if1}$. Only the absolute values of the frequencies make sense so that the power spectral density of $i_n^2(t)\cos^2\omega_{if1}t$ becomes $2ei_0$ for $0 < f < f_n - f_1$ and ei_0 for $f_n - f_1 < f < f_1 - f_1$ and also for $f_1 + f_1 < f < f_1 + f_n$ as plotted in Fig. 8.10a. Similarly the power spectral density that results from the term $i_n(t)\cos\omega_{if2}t$ becomes $2ei_0$ for $0 < f < f_n - f_2$ and ei_0 for $f_n - f_2 < f < f_2 - f_1$ and also for $f_2 + f_1 < f < f_2 + f_n$ as plotted in Fig. 8.10b. We arbitrarily assume $f_1 > f_2$. The sum of these two noise contributions is plotted in Fig. 8.10c. As seen from the figure the noise power that passes the narrow bandwidth B of the filter tuned at the frequency f_c depends on f_c. Since $f_c = f_1 - f_2$ is always smaller than $f_n - f_2$ we are dealing with two regions. The square of the output noise voltage for the bandwidth B within the frequency range of f_c is

(a) $0 < f_c < f_n - f_1$

$$\overline{v_n^2} = (2\gamma\beta AE_0)^2 \left(E_1^2 + E_2^2\right) 2ei_0 B = 32\gamma^2 ei_0^2 \left(i_1 + i_2\right) B \,, \tag{8.79}$$

(b) $f_n - f_1 < f_c < f_n - f_2$

$$\overline{v_n^2} = (2\gamma\beta AE_0)^2 \left(E_1^2 + 2E_2^2\right) ei_0 B = 16\gamma^2 ei_0^2 \left(i_1 + 2i_2\right) B \,. \tag{8.80}$$

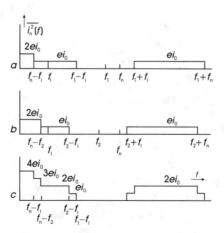

Fig. 8.10. Signal dependent spectral noise power. In parts (**a**) and (**b**) are shown the contributions related to the two heterodyne signals, respectively. Part (**c**) is the sum of (**a**) and (**b**)

The signal-to-noise ratio at the output of the bandpass filter for region (a), where the noise power is highest, becomes by using the (8.74), (8.78) and (8.79)

$$\frac{S}{N} = \frac{2i_1 i_2}{2e^2 B \left(\Delta f - f_c\right) + 8eB \left(i_1 + i_2\right)} \, . \tag{8.81}$$

For region (b) we find similarly

$$\frac{S}{N} = \frac{2i_1 i_2}{2e^2 B \left(\Delta f - f_c\right) + 4eB \left(i_1 + 2i_2\right)} \, . \tag{8.82}$$

It is seen that region (a) gives the largest NEP-value. Its minimum is found by taking the partial derivatives of i_1 and i_2 equal to zero. This gives $i_1 = i_2$ or equal received powers in the two signal beams. Taking $i_1 = i_2 = \frac{1}{2} i_s$ we get for region (a)

$$\frac{S}{N} = \frac{i_s^2}{4e^2 B \left(\Delta f - f_c\right) + 16eBi_s} \, . \tag{8.83}$$

From this we find the NEP by taking $S/N = 1$ and substituting $i_s = e\eta P_s / h\nu$. We get

$$\text{NEP} = \frac{h\nu}{e\eta} \left[8eB + \left\{ (8eB)^2 + 4e^2 B \left(\Delta f - f_c\right) \right\}^{1/2} \right] \, . \tag{8.84}$$

Since $f_c = f_1 - f_2$ is accurately known the bandwidth B can be taken very small so that $\Delta f - f_c$ is much larger than B. We may simplify (8.84) and get

$$\text{NEP} = \frac{h\nu}{e\eta} \left[4e^2 B \left(\Delta f - f_c\right) \right]^{1/2} \, . \tag{8.85}$$

We now proceed with the most conservative assumption that f_c is much smaller than Δf so that we neglect f_c in the square root. We finally find

$$\text{NEP} = \frac{h\nu \Delta f}{\eta} \left(\frac{4B}{\Delta f} \right)^{1/2} \, . \tag{8.86}$$

In case conventional heterodyne detection technique was applied the electronic bandwidth should have been Δf because of the Doppler uncertainty. Its NEP was then found as $h\nu(\Delta f)/\eta$. Thus by choosing B much smaller than Δf the NEP of the dual signal beam system is reduced by the factor $(4B/\Delta f)^{1/2}$. Finally we remark that the NEP in region (b), under the assumption $\Delta f \gg B$, will be identical and is also given by (8.86).

It should be noted that for large signal powers when $16eBi_s \gg 4e^2 B \Delta f$ or $P_s \gg (1/4) \left(h\nu \Delta f / \eta\right)$ we get according to (8.83)

$$\frac{S}{N} = \frac{\eta P_s}{16 h\nu B} \, , \tag{8.87}$$

which is apart from a factor $(1/16)$ the same as obtained for a conventional heterodyne system with a narrow bandwidth B.

8.7 Dual Signal Heterodyne Lock-In Amplification

The NEP of the system described Sect. 8.6 as given by (8.86) is limited by the bandwidth B of the narrow bandpass filter. In practice B is not much smaller than a few percent of the tuned frequency. If for instance f_c is of the order of 10 MHz the bandwidth is of the order of 100 kHz which still has a large bearing on the NEP.

A further reduction of the bandwidth, however, can be obtained by applying lock-in amplification, see Sect. 7.2. The output signal of the square-law device is after passing the blocking capacitor correlated with a reference signal of the same known frequency f_c as schematically shown in Fig. 8.11. This correlation technique is discussed in Chap. 7.

If the amplitude of one of the transmitted beams (the carrier wave), say E_1, varies slowly compared to the difference-frequency f_c this amplitude modulation is transferred to the mixed wave with frequency f_c at the output of the square-law device. The next step is then to demodulate this mixed wave. This can be done by rectifying the mixed wave with a double-pole reversing switch driven by an actuator which is synchronized with the frequency f_c as indicated in Fig. 8.11. The switching is such that for every polarity reversal of the mixed wave there is one of the switch. The effect of switching is in fact the multiplication of the mixed wave with a square wave of zero average value and amplitude one. Only the component with the frequency f_c is rectified by the correlation. All other signal components disappear. Because the phase of the mixed wave is unknown there will be a phase difference ϕ between the zero crossings of the two correlated waves so that the rectified voltage signal after passing the switch will be

$$V = \frac{2}{\pi}\gamma\beta^2 A^2 E_0^2 E_1\left(t\right) E_2 \cos\phi. \tag{8.88}$$

The factor $2/\pi$ comes from the integration of the rectified cosine-function.

The phase uncertainty of the lock-in amplification can be eliminated by applying simultaneously a second identical rectifying process of which the polarity switch is 90° phase shifted with respect to the first one. The corresponding output voltage is then

$$V = \frac{2}{\pi}\gamma\beta^2 A^2 E_0^2 E_1\left(t\right) E_2 \sin\phi. \tag{8.89}$$

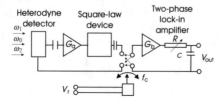

Fig. 8.11. Block diagram of a dual signal heterodyne lock-in amplifier system. After passing the square-law device and the capacitor the signal wave is rectified with a double-pole reversing switch driven by an actuator operating at the difference frequency of the two heterodyne signals

Taking the square root of the sum of the squares of the two output voltages given by the last two equations we have a phase-insensitive result. The voltage signal after passing the buffer amplifier G_b is led to the low-pass filter formed by the RC-circuit. The square of the signal voltage of the capacitor is

$$V_s^2 = \frac{64}{\pi^2} G_b^2 \gamma^2 i_0^2 i_2 i_1(t) , \qquad (8.90)$$

where we have substituted $i_1(t) = \frac{1}{2}\beta A E_1^2(t)$, $i_2 = \frac{1}{2}\beta A E_2^2$, and $i_0 = \frac{1}{2}\beta A E_0^2$.

The difference-frequency of the received beams f_c' differs relatively slightly from the transmitted difference-frequency f_c because of the Doppler shift as given by (8.64) so that the square wave frequency differs from that of the mixed wave. This results in a relatively slow time variation of ϕ. Since the phase is eliminated the final result is not affected by the frequency difference between f_c' and f_c.

The noise bandwidth of the low-pass filter is $(1/4RC)$. The square of the noise voltage on the capacitor can the be obtained from (8.78) by substituting $B = (1/4RC)$ and $f_c = 0$. We obtain after amplification for the square value

$$\overline{v_n^2} = \frac{2 G_b^2 \gamma^2 e^2 i_0^2 \Delta f}{RC} . \qquad (8.91)$$

We obtain from (8.90) and (8.91) for the signal-to-noise ratio

$$\frac{S}{N} = \frac{32 RC i_2 i_1(t)}{\pi^2 e^2 \Delta f} , \qquad (8.92)$$

where the time constant RC is limited by the maximum modulation frequency of the carrier beam. For obtaining the NEP we substitute the optimized condition $i_1 = i_2 = (1/2)i_s$ into (8.92) and find

$$\text{NEP} = \frac{h\nu\Delta f}{\eta} \left(\frac{\pi^2}{8 RC \Delta f} \right)^{1/2} . \qquad (8.93)$$

Depending on the RC-value the lock-in amplification makes the detection very sensitive. This technique of lock-in amplification also allows the possibility to reduce the adjusted bandwidth Δf which is taken at first relatively large to be sure to cover the unknown Doppler shifts. Receiving signal response with this lock-in amplification it is then possible to reduce the detection bandwidth Δf and therewith the noise in such a manner that the heterodyne frequencies $(1/2\pi)\omega_{if1}$ and $(1/2\pi)\omega_{if2}$ still remain within the confined band. In this way the Doppler-shifted frequencies are determined within a narrow frequency range. The next step is to switch simultaneously to the conventional heterodyne detection to find the intermediate frequency, say $(1/2\pi)\omega_{if1}$ and from this the Doppler shift giving the velocity of the target.

The usual way to find the distance to the target is the time of flight technique. This will be discussed later in Sect. 8.8.3

8.8 Dual Signal Wave Analyzer

The dual signal detection system can also be extended to analyze a transmitted waveform. This is schematically indicated in Fig. 8.12. Let us consider a waveform containing specific information, for instance, the read-out of a camera, which is superimposed as a repetitive wave on one of the two signal beams (carrier beam) as power modulation. If the frequency of the waveform is much smaller than that of the mixed wave the demodulated signal discussed in Sect. 8.7 and given by (8.90) describes before reaching the low-pass filter in Fig. 8.11 the waveform of interest, because $i_1(t)$ is proportional to the modulated power $P_1(t)$ of the carrier beam.

The voltage signal given by (8.90) is now connected to a waveform analyzer, see Sect. 7.3.2, that slices the periodic waveform and stores each slice with its noise in one of the channels. Because the waveform of duration T is periodic and each slice of width τ is sampled in its own channel the accompanying noise is filtered out. After sufficient sampling periods the transmitted waveform will be recovered and can be processed to display the read-out of the camera.

The effective time constant of each channel is

$$T_{\text{eff}} = RCN_{\text{b}}, \tag{8.94}$$

where RC is the time constant of a channel and $N_{\text{b}} = (T/\tau)$ is the number of channels.

The signal-to-noise ratio is now directly obtained from (8.92) by substituting the effective time constant in stead of the RC time of the low-pass filter of the lock-in amplifier. We obtain

$$\frac{S}{N} = \frac{32RCN_{\text{b}}i_2 i_1(t)}{\pi^2 e^2 \Delta f}. \tag{8.95}$$

For obtaining the NEP we substitute the optimized condition $i_1 = i_2 = \frac{1}{2}i_{\text{s}}$ into (8.95) and find

$$\text{NEP} = \frac{h\nu\Delta f}{\eta}\left(\frac{\pi^2}{8RCN_{\text{b}}\Delta f}\right)^{1/2}. \tag{8.96}$$

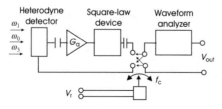

Fig. 8.12. Block diagram of a dual signal heterodyne waveform analyzer. After passing the square-law device and the capacitor the modulated signal wave with frequency equal to the difference-frequency of the two heterodyne signals is demodulated by a double-pole reversing switch driven by an actuator operating at the same difference-frequency of the two heterodyne signals. The demodulated signal wave is analyzed by the waveform analyzer

From the earlier analysis we can draw the following conclusions.

1. Although the noise equivalent bandwidth in the S/N-value given by (8.95) is $1/4RCN_b$ we remark that the sampling time of the analyzer is about $4RCN_b$ so that the sampling time for recovering the waveform is for a given noise bandwidth independent on the number N_b.
2. The effective sampling time of a channel is

$$T_{\text{sampling}} = 4RCN_b = 4R\frac{C}{\tau}T\,,\tag{8.97}$$

so that the capacitance of a channel is proportional to τ and inversely proportional to the number of channels. In other words the capacitance of a channel is inversely proportional to the resolution of the waveform. This is fortunate because miniaturization and increase of the number of channels to reach higher spatial resolution of the waveform is only feasible if the channel capacitance decreases with its number.

8.8.1 Space Communication

Communication with optical links between satellites in space has very attractive perspectives because directional laser beams with high-spectral power density, narrow line width, and low divergence can be sent from source to receiver. The received power varies as $1/\lambda^2$ so that the laser power requirement of the transmitter for space communication is strongly reduced compared with the use of microwaves. The reachable distance between emitter and receiver depends on the available laser power and the NEP of the detector. As we have seen before the NEP of incoherent detection is many orders of magnitude larger than in the case of coherent detection. Therefore it is attractive to develop the coherent detection for space communication. If varying or unknown Doppler shifts have to be taken into account the laser wavelength has a minimum as these shifts are proportional to the radiation frequency and the largest Doppler-shifted heterodyne frequency must fall within the frequency range of the detection system.

However, conventional heterodyne detection that must encompass the continuously changing Doppler frequency with unknown Doppler shifts requires a large bandwidth and moreover the modulated signal carrying the transmitted information will be seriously distorted due to the Doppler shifts. Demodulation may become very complex if possible at all. These problems can be overcome with dual signal heterodyne detection.

Let us consider the application of dual signal heterodyne detection for space communication in case the transmitter and receiver are moving relative to each other. The dual system provides as we have seen a mixed signal at the difference-frequency regardless of the Doppler shift. If the power of one of the transmitted beams is modulated according to the information it has to carry this modulation is transferred to the mixed signal at the receiver. The mixed signal beam can in fact be considered as a carrier wave with the difference-frequency having the same modulation as the transmitted wave. According to the scheme of Fig. 8.11 the modulated signal is fed into the lock-in amplifier where the reference signal is identical to the constant difference-frequency. The lock-in amplifier demodulates the mixed signal and the square of the signal voltage at the capacitor C is identical to the modulation of one of the transmitted beams provided that the chosen bandwidth $\frac{1}{4RC}$ is sufficient to follow the modulation. This is discussed in Sect. 7.2.

In the following we calculate the maximum distance in space for (audio) communication with a CO_2 laser operating at $10.6\,\mu m$ wavelength. We consider a modulation bandwidth $1/4RC$ equal to $10\,kHz$ and a relative velocity in the range of 0–$5\,km\,s^{-1}$ so that the variation of the Doppler shift for $\Delta v = 5\,km\,s^{-1}$ is $\Delta\nu = 2(\Delta v/c)\nu = 1\,GHz$. Applying (8.93) we find with $\eta = 0.8$ the NEP $= 1.75 \times 10^{-13}$ W.

The intensity at distance R becomes

$$I = \frac{P_1}{\Omega R^2} , \tag{8.98}$$

where $\Omega \approx (\lambda/d)^2$ is the beam divergence, d the aperture diameter of the transmitter, and P_1 the laser power. Having an aperture D for the receiver the signal power is then given by

$$\frac{P_s}{P_1} = \frac{\pi d^2 D^2}{4\lambda^2 R^2} . \tag{8.99}$$

Taking $P_1 = 10\,W$, $d = 20\,cm$, and $D = 100\,cm$ the distance to receive ten times the NEP becomes 4.2×10^7 km.

8.8.2 Transmitting Photographs

Next we consider the transmission of photographs from Jupiter at a distance of 7.8×10^8 km and their recovery by a waveform analyzer. We take again a CO_2 laser, $P_1 = 10\,W$, $d = 20\,cm$, $D = 100\,cm$ and $\Delta f = 1\,GHz$. The received signal power by using (8.99) is 5×10^{-15} W. Let we require a NEP of only 1% of the signal power or NEP $= 5 \times 10^{-17}$ W. Substituting this value into (8.96) we find $RCN_b = 300\,s$ so that the effective sampling time of the waveform to transmit a photograph is 20 min. The sampling time is inversely proportional to the square of the laser power. For a laser power of $100\,W$ the sampling time is $12\,s$, whereas the travel time of the signal from Jupiter is 43 min.

8.8.3 Laser Radar

As another example of the dual signal beam heterodyne detection we consider CO_2-laser radar as a rangefinder in space operating at $10.6\,\mu m$ wavelength. We specify the range R to be $1,000\,km$ and we require a range resolution of

1.5 km. The common type of range finder uses a time of flight technique to measure the range. This means that a laser pulse is transmitted toward a target and radiation scattered from the target is collected. The time delay τ between transmission of the laser pulse and the detection of the backscatter is used to calculate R according to

$$R = \frac{c\tau}{2} . \qquad (8.100)$$

The round-trip time for 1,000 km is about 7 ms. The range resolution of 1.5 km requires a detector rise time of $10\,\mu s$ i.e., the travel time for twice the range resolution which corresponds with an electronic bandwidth B of 100 kHz. The laser pulse is transmitted through a telescope with a large aperture diameter d of 20 cm to reduce the beam divergence $\Omega \approx (\lambda/d)^2$. The receiver channel collects the radiation scattered from the target and focuses this onto the detector. The larger the receiver area the more power will be detected. Usually the transmitted and received radiation have the same aperture.

The intensity I of the laser radiation at the target is

$$I = \frac{P_1}{\Omega R^2} , \qquad (8.101)$$

where P_1 is the transmitted laser power. The incident power P_{in} on the target with surface area A becomes $P_{in} = AI$. Next we estimate the power reflected from the target reaching the receiver. Obviously the surface finish and detailed shape of the target will influence the return. As these parameters are in general uncertain, the usual solution is to assume the reflection to be diffuse with a so-called Lambertian profile similar to thermal radiation emitted by a surface. According to (1.45) the power integrated over the narrow frequency band is

$$dP = \cos\theta L A \, d\Omega , \qquad (8.102)$$

where the radiance $L = B\delta\nu$ and $\delta\nu$ the bandwidth of the scattered radiation. The total power scattered by the surface A integrated over the solid angle Ω is assumed equal to rP_{in} where r is the reflectivity of the surface or

$$P = \pi A L = rP_{in} = \frac{rAP_1}{\Omega R^2} . \qquad (8.103)$$

For simplicity the receiver will be assumed to lie along a normal to the surface A. The solid angle of the receive channel at the distance R is

$$\Omega_{rec} = \frac{\pi d^2}{4R^2} . \qquad (8.104)$$

The return signal power P_s received by the detector is then

$$P_s = \frac{\pi d^2}{4R^2} L A . \qquad (8.105)$$

Substituting (8.103) into (8.105) we find

$$\frac{P_{\mathrm{s}}}{P_{\mathrm{l}}} = \frac{rAd^2}{4\Omega R^4} \tag{8.106}$$

Assuming a radar cross-section of $1\,\mathrm{m}^2$ and a reflectivity of 50% we find $P_{\mathrm{s}}/P_{\mathrm{l}} = 2 \times 10^{-18}$

Let us first investigate the required laser power for comparison in case we apply incoherent detection. For this purpose we choose a PbSnTe photodiode which is dark current limited with $D^* = 2 \times 10^{10}\,\mathrm{W}^{-1}\,\mathrm{cm}\,\mathrm{Hz}^{1/2}$. With $B = 100\,\mathrm{kHz}$ and choosing a detector area of $0.01\,\mathrm{mm}^2$ the $\mathrm{NEP} = \frac{(AB)^{1/2}}{D^*} = 1.5 \times 10^{-10}\,\mathrm{W}$. For $\frac{S}{N} = 1$ we need $\mathrm{NEP} \times \frac{P_{\mathrm{l}}}{P_{\mathrm{s}}} = 7.5 \times 10^7\,\mathrm{W}$.

With conventional heterodyne detection we have to adjust a bandwidth that is sufficiently large to catch the uncertain Doppler shift of the return signal. Covering a radial velocity in the range of $0\text{–}5\,\mathrm{km\,s}^{-1}$ the Doppler shift is $\Delta\nu = (2vc)\nu = 1\,\mathrm{GHz}$. Since high-frequency response is required we use a photodiode with a quantum efficiency of 80%. Thus with $B = 1\,\mathrm{GHz}$ and $\eta = 0.8$ we find $\mathrm{NEP} = (h\nu B/\eta) = 2.5 \times 10^{-11}\,\mathrm{W}$.

In case of dual signal beam we apply (8.86) with $\Delta f = 1\,\mathrm{GHz}$ and $B = 100\,\mathrm{kHz}$ and find $\mathrm{NEP} = 5 \times 10^{-13}\,\mathrm{W}$. The laser power to reach $(S/N) = 1$ becomes now $2.5 \times 10^5\,\mathrm{W}$. For pulses of $10\,\mu\mathrm{s}$ the pulse energy for the two lines together is $2.5\,\mathrm{J}$. The specifications of the laser are a pulsed system with pulses of $10\,\mu\mathrm{s}$ containing two lines with a fixed frequency difference f_{c} of, say, $10\,\mathrm{MHz}$. The Doppler shift of this difference-frequency is about $700\,\mathrm{Hz}$ which is much smaller than the required line width of $100\,\mathrm{kHz}$. The repetition rate of the laser depends on the available power. For instance with $10\,\mathrm{Hz}$ we need an average laser power of $25\,\mathrm{W}$. To get the signal well above the noise level the laser pulses must be at least $10\,\mathrm{J}$ and the average laser power in that case is at least $100\,\mathrm{W}$.

The target velocity can of course be found from the beat frequency between the frequencies of the local oscillator and one laser line. The heterodyne detection is then conventional for which the required power, as we have seen, is much higher and probably not feasible. However, the measured range by each pulse as a function of time yields the target velocity in the straight forward way.

9

Fast Detection of Weak and Noisy Signals

Weak signals buried in noise can be resolved by applying correlation techniques as was pointed out in Chap. 6 and 7. It was seen that, for instance, by using the lock-in amplifier the noise-equivalent-power(NEP) decreases proportionally to the sampling time RC of the low-pass filter. As a consequence of sampling the systems become slow and the maximum detectable amplitude frequency of the signal is $1/RC$.

A very different technique that provides fast response of the incident photons of low power beams buried in noise is based on the typical multiplication mechanism of fast photomultipliers or in some cases of avalanche photodiodes. We assume hereby that the average separation time between photons is larger than the time response of the detector. As discussed already in Chap. 4 an important feature of the photomultiplier is the ability to detect a single photon. Its multiplication process reduces the relative contribution of the amplifier noise compared to other noises so that the system may reach signal noise limitation in the time domain of the single photon.

The inherent problem of detecting the single photons with fast response is the large electronic bandwidth of the detection system, typical in the order of GHz, which has a large bearing on the noise signals. Hence the Johnson or amplifier noise is much larger than in other fast detection systems with narrower bandwidth.

However, the fast detection of the individual photons of a low power beam does not mean fast information transfer concerning beam intensity and its modulation. Because of the inherent statistical spread in time of the photoelectrons reliable information requires a minimum number of photons. The ultimate frequency for meaningful signal response of low power beams is therefore, as we shall see, proportional to the beam power.

9.1 Suppressing Amplifier Noise with Detection Discriminator

Let us consider the detection of $0.5\,\mu\mathrm{m}$ radiation with an average power $10^{-12}\,\mathrm{W}$. The used photomultiplier tube has the following specifications: the current gain $G = 10^7$, output impedance $R = 100\,\Omega$, quantum efficiency $\eta = 0.15$, and bandwidth $B = 1\,\mathrm{GHz}$. A single photoelectron has a duration $\tau_e = \frac{1}{2B} = 0.5\,\mathrm{ns}$. The average separation time between two photons is $\tau_{\mathrm{sep}} = \frac{h\nu}{\eta P_s} = 2.6 \times 10^{-6}\,\mathrm{s}$ which is 5,200 times the pulse duration. The noise consists partly of photoelectrons emitted through the absorption of background photons or emitted as the result of thermal emission from cathode and dynodes (dark current). These noise photoelectrons can be considerably reduced by shielding the detector from the outside and by cooling at cryogenic temperature. Let us assume that these noise pulses are negligible and that we are only dealing with the amplifier or Johnson noise with an average square value $\overline{i_n^2} = 4kT_{\mathrm{eff}}B/R = 3.4 \times 10^{-13}\,\mathrm{A}^2$. Because of the large bandwidth of the photomultiplier the Johnson noise has strong fluctuations. The average signal current is $i_s = \frac{e\eta P_s G}{h\nu} = 6 \times 10^{-7}\,\mathrm{A}$. The signal-to-noise ratio is then $i_s^2/\overline{i_n^2} \approx 1$. Detection in a straightforward way (without correlation) is thus impossible.

However the peak current of a single photoelectron is $i_e = \frac{eG}{\tau_e} = 3.2 \times 10^{-3}\,\mathrm{A}$ is much larger than the average amplifier noise current $\sqrt{\overline{i_n^2}} = 5.8 \times 10^{-7}\,\mathrm{A}$ so that the single photons are observed. The unwanted Johnson noise can be considerably reduced when the photomultiplier pulses are passed to an operational amplifier with feedback resistor R_f and then to a discriminator which rejects all pulses below a certain voltage level. If the threshold voltage $v_T = R_f i_T$ is properly chosen, a great deal of the noise will be eliminated whereas only a very small part of the signal pulses are rejected. In the following we shall discuss the signal-to-noise ratio of the filtered pulses and the minimum sampling time for getting a meaningful signal response.

The random Johnson and amplifier noise resulting from a large number of independent producers can be described with a Gaussian current probability density as given in the Appendix A.2.3. The part of the noise power $\overline{i_{nT}^2}$ that passes the threshold current i_T is

$$\overline{i_{nT}^2} = \frac{1}{\sqrt{2\pi \overline{i_n^2}}} \int_{i_T}^{\infty} i_n^2 e^{-i_n^2/2\overline{i_n^2}} \mathrm{d}i_n = \alpha\left(\frac{i_T}{\sqrt{\overline{i_n^2}}}\right)\overline{i_n^2}, \tag{9.1}$$

where

$$\alpha\left(\frac{i_T}{\sqrt{\overline{i_n^2}}}\right) = \frac{2}{\sqrt{\pi}} \int_{i_T/\sqrt{2\overline{i_n^2}}}^{\infty} x^2 e^{-x^2} \mathrm{d}x. \tag{9.2}$$

The value of α is smaller than one and decreases very fast with increasing $i_T/\sqrt{\overline{i_n^2}}$ as can be seen in Fig. 9.1.

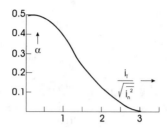

Fig. 9.1. Noise reduction depending on the threshold current

The bandwidth B of the system determines the duration of any fluctuation of the detected current. The shortest noise pulses are as long as a photoelectron pulse. The fluctuations of the noise current are added to the photoelectron pulses so that the statistical distribution $P_e(i)$ of the signal current of photoelectrons has the probability density

$$P_e(i) = \frac{1}{\sqrt{2\pi \overline{i_n^2}}} e^{-(i-i_e)^2/2\overline{i_n^2}}. \tag{9.3}$$

The observed average photoelectron current is

$$i_{eT} = \frac{1}{\sqrt{2\pi \overline{i_n^2}}} \int_{i_T}^{\infty} i e^{-(i-i_e)^2/2\overline{i_n^2}} di \tag{9.4}$$

or

$$i_{eT} = \frac{i_e}{\sqrt{\pi}} \int_{-\frac{(i_e - i_T)}{\sqrt{2\overline{i_n^2}}}}^{\infty} e^{-x^2} dx + \sqrt{\frac{2\overline{i_n^2}}{\pi}} \int_{-\frac{(i_e - i_T)}{\sqrt{2\overline{i_n^2}}}}^{\infty} x e^{-x^2} dx. \tag{9.5}$$

The second integral of (9.5) represents the statistical contribution of the instantaneous noise current which is for $(i_e - i_T) > \sqrt{2\overline{i_n^2}}$ negligible compared to the first term. Usually $(i_e - i_T) \gg \sqrt{2\overline{i_n^2}}$ so that we have

$$i_{eT} = i_e. \tag{9.6}$$

Also the observed pulse amplitudes fluctuate because of the addition of noise current. These fluctuations associated with $P_e(i)$ give a noise contribution equal to

$$\overline{i_{ne}^2} = \overline{(i - i_e)^2} = \frac{1}{\sqrt{2\pi \overline{i_n^2}}} \int_{i_T}^{\infty} (i - i_e)^2 e^{-(i-i_e)^2/2\overline{i_n^2}} di \tag{9.7}$$

or

$$\overline{i_{ne}^2} = \frac{2}{\sqrt{\pi}} \overline{i_n^2} \int_{-(i_e - i_T)/\sqrt{2\overline{i_n^2}}}^{\infty} x^2 e^{-x^2} dx, \tag{9.8}$$

which is for $i_e - i_T/\sqrt{2\overline{i_n^2}} \gg 1$ equal to $\overline{i_n^2}$. Thus the signal-to-noise ratio of the observed current pulses becomes $i_e^2/\overline{i_n^2} = 3 \times 10^7$ so that the noise of the observed pulse amplitudes is negligible. Thus with the use of a discriminator the amplifier noise is practically eliminated. The next question is then how much noise is contributed by the signal itself. This noise results from the fluctuating time intervals between the individual micropulses. Even when the signal beam is delivered by a coherent source (laser) with sufficient long temporal coherence so that its intensity is constant over the observation period, the time distribution of the emitted photoelectrons obey the Poisson statistics. This is derived in Appendix A.2.4 on the basis of the plausible assumption that the probability of emitting an electron is time independent for a constant incident beam intensity. Thus the number n of photoelectrons fluctuates during a period T according to $\overline{\Delta n^2} = \overline{(n - n_s)^2} = n_s$, where n_s is the average value. We obtain for the signal limitation

$$\frac{S}{N} = \frac{n_s^2}{\overline{\Delta n^2}} = n_s = \frac{\eta P_s T}{h\nu} , \qquad (9.9)$$

where the observation time T is set by the sampling time of, for instance, a low-pass filter connected to the detection system. In our example we find $S/N = 4 \times 10^5\, T$. For obtaining $S/N = 100$ the low-pass RC-filter must have a frequency bandwidth $\frac{1}{4RC} = 2 \times 10^3\,\mathrm{Hz}$.

It should be noted that the S/N-ratio given by (9.9) is the ultimate performance for measuring the beam intensity and its modulation. For useful information a sufficiently large S/N-ratio is required which is reached after an appropriate sampling time. In other words, the ultimate frequency for meaningful signal response of a low power beam is proportional to the average beam power.

In the above example we have neglected the disturbing background and dark current signals. This is not always realistic. However, the micropulses from the background and dark current can be eliminated by applying lock-in amplification discussed in Sect. 7.2. Then by chopping the signal beam the dark and background signals will be eliminated and only their fluctuations remain but decrease with the sampling time. This technique will be discussed in Sect. 9.2.

9.2 Photon Counting

Instead of measuring the current it is from the electronic engineering point of view more attractive to count the photoelectron pulses. In this technique the photomultiplier pulses are passed to an operational amplifier with feedback resistance R_f and then to a discriminator which rejects all pulses below a certain height. The transmitted pulses are counted by a digital counter. The amplifier is used to reach the voltage $i_e R_f = 2BeGR_f$ that is compatible with the operating threshold voltage of the discriminator in the range of $10\,\mathrm{mV}$–$1\,\mathrm{V}$. The amplifier must have a fast rise time with low input noise and good linearity. As long as the smaller pulses that originate from the thermionic emission of the dynodes are rejected, the stability of the photomultiplier tube is not relevant in this technique. On the other hand noise pulses that just pass the discriminator give full counts.

Since a photomultiplier has a low dark current arising from spontaneous electron emission and a strong build-in amplification system which discriminates Johnson and additional amplifier noise, the device is extremely sensitive to select single ultraviolet, visible, or near infrared photons. In spite of the fact that the Johnson or amplifier noise has, because of the large bandwidth B of the photomultiplier, relatively strong noise pulses, the accompanying noise can for a great deal be rejected electronically from counting. This approach is, as we have seen in Sect. 9.1, only relevant in case of very low power so that the average time interval between photons is much greater than the width of a photoelectron pulse and when the photoelectron pulse current is much larger than the average amplifier noise current. Although in practice the chosen threshold current $i_T = v_T/R_f$ of the discriminator is much larger than the root mean square noise current $\sqrt{\overline{i_n^2}}$ of the amplifier or Johnson noise there remain probabilities that noise pulses pass the discriminator and mix with the signal counts. This happens to that part of the Johnson noise current greater than i_T. Often the discriminator has also an upper threshold current level i_U to reject pulses with currents larger than i_U. These high noise current pulses may be due to high energy photons generating multiple photoelectrons like cosmic rays coming from outer space.

The minimum duration of noise fluctuations is equal to the time constant of the photomultiplier $\tau = 1/2B$. The rate of noise pulses is B. Assuming a Gaussian probability density of the noise current the probability to register a noise pulse is given by

$$P_n = \frac{1}{\sqrt{2\pi \overline{i_n^2}}} \int_{i_T}^{\infty} e^{-i_n^2/2\overline{i_n^2}} \, di_n = \frac{1}{\sqrt{\pi}} \int_{i_T/\sqrt{2\overline{i_n^2}}}^{\infty} e^{-x^2} \, dx$$

or

$$P_n = \frac{1}{2} \left[1 - \mathrm{erf}\left(i_T/\sqrt{2\overline{i_n^2}} \right) \right], \tag{9.10}$$

where it is assumed that the upper threshold value $i_U \gg \sqrt{\overline{i_n^2}}$ so that the upper integration limit i_U can be replaced by ∞. The rate N_n of detected noise counts becomes

$$N_n = P_n B. \tag{9.11}$$

A photon count occurs when the sum of photon and noise current is above i_T. The detection probability P_d of a single photoelectron is then given by

$$P_d = 1 - \frac{1}{\sqrt{\pi}} \int_{(i_e - i_T)/\sqrt{2\overline{i_n^2}}}^{\infty} e^{-x^2} \, dx \tag{9.12}$$

or

$$P_{\mathrm{d}} = \frac{1}{2}\left[1 + \mathrm{erf}\left(\frac{i_{\mathrm{e}} - i_{\mathrm{T}}}{\sqrt{2i_{\mathrm{n}}^2}}\right)\right] , \qquad (9.13)$$

where i_{e} is the current of a single photoelectron. In case $(i_{\mathrm{e}} - i_{\mathrm{T}})$ is a few times $\sqrt{i_{\mathrm{n}}^2}$ so that the second term of (9.12) is negligible we get $P_{\mathrm{d}} = 1$.

Next we raise the question to what extent pulse counting gives reliable information on the beam intensity and its modulation. To answer this question let us first consider the ultimate beam quality with signal limitation i.e., signal noise is much larger than all other noise signals. In that case we have $S/N = N_{\mathrm{s}}\tau = \eta P_{\mathrm{s}}\tau/h\nu$. For $S/N = 1$ the signal pulse rate is $N_{\mathrm{s}} = 1/\tau$ which means i_{e} is equal to the average current i_{s} and the average time distance between two photoelectrons is equal to the pulse duration. Discrimination by eliminating a part of the Johnson noise is not possible. In other words for $S/N = 1$ photon counting does not make sense. Thus if, however, the beam intensity is so low that photon counting can be applied we must extend the observation time period for obtaining reliable information. For obtaining $S/N = 1$ we find for the noise equivalent sampling time

$$T_{\mathrm{NE}} = \frac{i_{\mathrm{e}}}{i_{\mathrm{s}}}\tau . \qquad (9.14)$$

The maximum frequency for obtaining reliable information from the signal beam is proportional to $\frac{1}{T_{\mathrm{NE}}}$. Thus the larger the signal pulse rate and the more they are free from noise signals the larger is the useful frequency of the signal response.

In case the background photoelectrons, dark current pulses, and the remaining Johnson noise that are passed to the discriminator are not negligible, further improvement of the signal-to-noise ratio can be obtained with lock-in amplification. Following the treatment given in Sect. 7.2 we note that by chopping the signal beam only the signal is lock-in amplified and the background, dark current, and Johnson noise will vanish except for their fluctuations or their shot noise. Exactly the same can be obtained by counting the pulses [35]. During the period the beam is on the photomultiplier, the counts correspond to the signal plus noise. During the blocked period the counts correspond only to noise pulses. Substracting the two counts yields the signal counts plus the fluctuations of the counts. These random fluctuations are not abstracted but added.

The signal current is for a chopper with duty cycle of 50% given by $i_{\mathrm{s}} = \frac{1}{2}eN_{\mathrm{s}}$. The spectral noise power density of the chopped signal is $2ei_{\mathrm{s}} = e^2 N_{\mathrm{s}}$. The spectral noise power density of the other noise sources which are not interrupted by the chopper is $2e^2$ times the pulse rates. The bandwidth of the noise power is $1/2T$ where T is the sampling time. Taking the signal-to-noise ratio we get

$$\frac{S}{N} = \frac{T}{2}\frac{N_{\mathrm{s}}^2}{N_{\mathrm{s}} + 2N_{\mathrm{b}} + 2N_{\mathrm{d}} + 2BP_{\mathrm{n}}} , \qquad (9.15)$$

where N_s, N_b, and N_d are the signal, background, and dark current pulse rates, respectively.

The noise can also be calculated by considering the Poisson statistics of the photoelectrons. Dealing with the sampling time T we have a total average number of $\bar{n} = T(\frac{1}{2}N_s + N_b + N_d + BP_n)$ photoelectrons. According to Poisson statistics we have $\overline{\Delta n^2} = \bar{n}$. The counted signal pulses are equal to $n_s = \frac{1}{2}TN_s$. The signal-to-noise ratio is equal to $n_s^2/\overline{\Delta n^2}$ which gives after substituting the values for n_s and $\overline{\Delta n^2}$ again the result of (9.15).

Usually with sufficient discrimination N_d and BP_n are negligible compared with N_b so that the system is background limited. The noise equivalent count rate NEC becomes

$$\text{NEC}_{\text{BL}} = \frac{1}{T}\left[1 + (1 + 4TN_b)^{1/2}\right].\tag{9.16}$$

If $N_b \gg \frac{1}{T}$ we find

$$\text{NEC}_{\text{BL}} = 2\sqrt{\frac{N_b}{T}}\tag{9.17}$$

with the condition that NEC_{BL} is considerably smaller than $2B$ in order to make photon counting possible. In case the number of background signal pulses is also negligible we reach the ultimate signal limitation and obtain for the noise equivalent count rate

$$\text{NEC}_{\text{SL}} = \frac{2}{T},\tag{9.18}$$

which of course must be larger than NEC_{BL}. The factor 2 arises from the 50% duty cycle of the chopper. The corresponding NEP becomes

$$\text{NEP}_{\text{SL}} = \frac{2h\nu_s}{\eta T}.\tag{9.19}$$

A

Appendix

A.1 Microcurrent Pulse

When an electron is emitted from the cathode and travels to the anode a current $i(t)$ flows in the external leads connecting the electrodes. The continuous current during the travel of the emitted electron may be understood as the continuous rate of change of the image charges on the two electrodes that has to be supplied by $i(t)$. At the moment of emission an equal and opposite image charge at the cathode starts the current flow $i(t)$. The image charge at the cathode decreases, whereas the corresponding image charge at the anode increases as the electron travels away from the cathode. When the electron arrives at the anode it encounters an equal and opposite charge with which it is neutralized upon impact.

To evaluate this process we consider the trip of an electron from the cathode to the anode as illustrated in Fig. A.1. While the electron is moving to the anode due to the applied potential by the battery there is continuity of current which means that at any position of the circuit the current is equal to that between the electrodes. As long as the electron is moving there is current with the value

$$i(t) = \frac{ev(t)}{d} \, , \tag{A.1}$$

where $v(t)$ and d are the velocity and the distance between the electrodes, respectively. Integrating $i(t)$ over the transit time τ gives the charge of an electron. After arrival at the anode the current of this single electron becomes zero. Thus each individual released electron generates a micropulse in the external circuit with a duration equal to the transit time.

This process of external current flow is also the case for an electron–hole pair that is generated in a semiconductor. The charge carriers travel to the collectors; the holes to the negative and the electrons to the positive electrode. A current pulse of an electron–hole pair can then be considered as the sum of two pulses generated by the individual carriers. Although the differences of

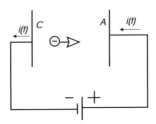

Fig. A.1. Microcurrent pulse circuit

velocities of the carriers and the differences of distances to the electrodes result
in different pulse durations for the two carriers the external pulse duration
is equal to the longest of the two pulses. Each generated electron–hole pair
creates in the absence of electrode emission a current pulse with a charge of
one electron in the external circuit.

A.2 Statistics

A.2.1 Binomial Distribution

We derive the probability of occurrence of n random events like the emission of
photons or electrons by a cathode in a random process so that the probability
of occurrence in a time interval or space area is independent of the probability of occurrence at any other time interval or space area.

Let p be the probability that a certain experiment is successful and $q =
1 - p$ the probability of failure. If the experiment takes place m times there is
a series of successes and failures. The total probability is the product of the
individual probabilities or $ppqpqq \cdots p$. The series contains n successes and
$m - n$ failures so that for a particular sequence of events its probability of n
successes is given by $p^n q^{m-n}$. This probability of successes can be found for
a number of different sequences which all contain n successes. This number is
equal to $m!/n!(m-n)!$. Thus the probability of finding n successes and $m - n$
failures is

$$P_m(n) = \frac{m!}{n!(m-n)!} p^n q^{m-n} , \tag{A.2}$$

which is called the binomial distribution because $P_m(n)$ is the $(n+1)$th term
in the binomial expansion of $(p + q)^m$. This distribution is, as expected, nor-
malized because

$$\sum_{n=0}^{m} P_m(n) = \sum_{n=0}^{m} \frac{m!}{n!(m-n)!} p^n q^{m-n} = (p + q)^m = 1 . \tag{A.3}$$

The average value $\bar{n}$, obviously equal to pm, is given by

$$\overline{n} = \sum_{n=1}^{m} n P_m(n) = \sum_{n=1}^{m} \frac{m!}{(n-1)!(m-n)!} p^n q^{m-n}$$

$$= m \sum_{n=1}^{m} \frac{(m-1)!}{(n-1)!(m-n)!} p^n q^{m-n}$$

$$= mp(p+q)^{m-1} = mp \qquad (A.4)$$

because $q = 1 - p$

The value $\overline{n^2}$ can be derived by

$$\overline{n^2} = \sum_{n=1}^{m} n^2 P_m(n) = \sum_{n=2}^{m} n(n-1) P_m(n) + \sum_{n=1}^{m} n P_m(n)$$

$$= \sum_{n=2}^{m} \frac{m!}{(n-2)!(m-n)!} p^n q^{m-n} + mp$$

$$= p^2 m(m-1) \sum_{n=2}^{m} \frac{(m-2)!}{(n-2)!(m-n)!} p^{n-2} q^{m-n} + mp$$

$$= p^2 m(m-1)(p+q)^{m-2} + mp = \overline{n}^2 + \overline{n}(1-p),$$

where we substituted $(pm)^2 = \overline{n}^2$ and $q = 1 - p$

Since $\overline{\Delta n^2} = \overline{(n - \overline{n})^2} = \overline{n^2} - \overline{n}^2$ we have

$$\overline{\Delta n^2} = \overline{n}(1-p), \qquad (A.5)$$

which does not depend on m.

If $p \ll 1$ we may write for a binomial distribution

$$\overline{\Delta n^2} = \overline{n}. \qquad (A.6)$$

A.2.2 Poisson Distribution

The probability function of (A.2) becomes a Poisson distribution if m tends to infinity and p tends to zero, while the product $mp = \overline{n}$ remains finite. This can be derived by writing (A.2) as

$$\frac{m!}{n!(m-n)!} p^n (1-p)^{m-n} = \frac{(mp)^n}{n!} \left(1 - \frac{mp}{m}\right)^m \frac{m!}{m^n(m-n)!} (1-p)^{-n}. \quad (A.7)$$

Substitute the limiting values of

$$\lim_{p \to 0} (1-p)^{-n} = 1,$$

$$\lim_{m \to \infty} \frac{m!}{m^n(m-n)!} = \lim_{m \to \infty} \frac{m(m-1)\dots(m-n+1)}{m^n} = 1$$

and

$$\lim_{m\to\infty}\left(1-\frac{mp}{m}\right)^m = \lim_{m\to\infty}\left(1-\frac{\bar{n}}{m}\right)^m = e^{-\bar{n}}.$$

We obtain for $mp \to \bar{n}$

$$\lim_{m\to\infty}\frac{m!}{n!(m-n)!}p^n(1-p)^{m-n} = \frac{\bar{n}^n}{n!}e^{-\bar{n}}. \tag{A.8}$$

Thus the Poisson distribution is given by

$$P(n) = \frac{\bar{n}^n}{n!}e^{-\bar{n}}. \tag{A.9}$$

It is seen that the Poisson statistics contains neither m nor p and can be applied to evaluate statistical processes where both m and p are meaningless. Since $p \to 0$ we find as already shown in (A.6) that for the Poisson distribution holds

$$\overline{\Delta n^2} = \bar{n}. \tag{A.10}$$

A.2.3 Gaussian Distribution

The Poisson distribution has for large values of $\bar{n}$ a sharp maximum with a Gaussian distribution around this maximum. This can be seen as follows. Dealing with large values of n we apply Stirling's asymptotic formula for the factorial of a large number given by

$$n! = \sqrt{2\pi n}\, n^n e^{-n}. \tag{A.11}$$

Substituting this formula into (A.9) we obtain for $\log P(n)$

$$\log P(n) = n\log\bar{n} - \bar{n} - \frac{1}{2}\log(2\pi) - \left(n+\frac{1}{2}\right)\log n + n. \tag{A.12}$$

We now substitute $(n - \bar{n}) = \delta$ and expand the right hand side of (A.12) into a Tayler series for $\delta/\bar{n}$. We get

$$\log P(n) = -\frac{1}{2}\bar{n}\left(\frac{\delta}{\bar{n}}\right)^2 - \frac{1}{2}\left(\frac{\delta}{\bar{n}}\right) - \frac{1}{2}\log\left(2\pi\bar{n}\right) \tag{A.13}$$

plus higher order terms in $\delta/\bar{n}$ which are neglected. It is seen that except for $\delta = 0$ or 1 the first term becomes much larger than the second one so that the second term may be neglected. We then obtain

$$P(n) = \frac{e^{-\delta^2/(2\bar{n})}}{\sqrt{2\pi\bar{n}}} = \frac{e^{-(n-\bar{n})^2/(2\bar{n})}}{\sqrt{2\pi\bar{n}}}, \tag{A.14}$$

which is the Gaussian distribution with a sharp maximum for $n = \bar{n}$ if $\bar{n}$ is large. Since for a Poisson distribution $\overline{(n - \bar{n})^2} = \overline{\Delta n^2} = \bar{n}$, it is not surprising to obtain it also for this Gaussian distribution.

Often the number of photon electrons which obey the Poisson statistics is very large during the considered time interval $\tau = 1/2B$ so that it is permissible to replace the Poisson distribution by the Gaussian one.

In general, white noise like shot noise and Johnson noise result from the concerted action of a large number of independent producers. The statistical distribution of the noise current is therefore Gaussian. The probability density of this noise current i_n is given by

$$F\left(i_\mathrm{n}\right) = \frac{1}{\sqrt{2\pi \overline{i_\mathrm{n}^2}}} \mathrm{e}^{-i_\mathrm{n}^2/2\overline{i_\mathrm{n}^2}} \tag{A.15}$$

so that the average square value of the noise current becomes

$$\int_{-\infty}^{\infty} i_\mathrm{n}^2 F\left(i_\mathrm{n}\right) \mathrm{d}i_\mathrm{n} = \overline{i_\mathrm{n}^2}. \tag{A.16}$$

A.2.4 Photoelectron Statistics

Although photoemission is a quantum mechanical process we may consider the statistics of the photoelectrons classically by making the plausible assumption that the probability of creating a photoelectron is proportional to the incident radiation power. We shall now derive the statistics of the photoelectrons for constant incident power on the photocathode. This implies temporal coherence of the incident radiation source during the observation time. Such an assumption applies mainly to laser sources.

The probability Δp that a photoelectron is created in a small time interval Δt at the time t when the photocathode with efficiency η is illuminated with the constant radiation power P_s is

$$\Delta p = \frac{\eta P_\mathrm{s}}{h\nu} \Delta t. \tag{A.17}$$

If there are no random fluctuations in the power P_s, then we assume that the probabilities of producing electrons in equal distinct time intervals are statistically independent. The probability that no photoelectron is created in the time interval Δt is $1 - \Delta p$. If we now consider many successive time intervals Δt_n during the entire observation time T there is a series of successful events and of failures. For mathematical reasons we split up the observation time T into a large number m of successive small time intervals Δt. The total probability is the product of all infinite small probabilities during the period T. For n successful events and $m - n$ failures we get the total probability equal to $(\Delta p)^n (1-\Delta p)^{m-n}$. The probability distribution of n successful events within m small time intervals $\mathrm{d}t$ can be arranged in may different ways. This number is equal to $m!/n!(m-n)!$. Thus the total probability to create n photo-electrons is

$$P_m(n) = \frac{m!}{n!(m-n)!}(\Delta p)^n(1-\Delta p)^{m-n}. \qquad (A.18)$$

If m goes to infinity and Δp to zero, while the product $m\Delta p = \bar{n}$ is equal to the average number of successfully created photoelectrons, the further evaluation of $P_m(n)$ results in a Poisson distribution as discussed in Appendix A.2.2. Thus the statistics of photoelectrons created by a constant incident power on a photocathode obeys the Poisson distribution.

A.3 Multiplication Factor M_{n}

Solving the quasisteady-state rate equations for the avalanche multiplication

$$\frac{J_{\mathrm{n}}(x)}{\mathrm{d}x} = \alpha J_{\mathrm{n}}(x) + \beta J_{\mathrm{p}}(x) \qquad (A.19)$$

and

$$-\frac{J_{\mathrm{p}}(x)}{\mathrm{d}x} = \alpha J_{\mathrm{n}}(x) + \beta J_{\mathrm{p}}(x), \qquad (A.20)$$

which yields

$$J_{\mathrm{n}}(x) + J_{\mathrm{p}}(x) = J = \text{const}. \qquad (A.21)$$

We substitute (A.21) into (A.20) and obtain

$$\frac{J_{\mathrm{p}}(x)}{\mathrm{d}x} = (\alpha - \beta)J_{\mathrm{p}}(x) - \alpha J. \qquad (A.22)$$

Next we substitute $J_{\mathrm{p}}(x) = q(x)r(x)$ into (A.22) and obtain

$$q\frac{\mathrm{d}r}{\mathrm{d}x} + r\frac{\mathrm{d}q}{\mathrm{d}x} = (\alpha - \beta)qr - \alpha J. \qquad (A.23)$$

Solving for $r(x)$ by equating $\mathrm{d}r/\mathrm{d}x = (\alpha - \beta)r$ yields

$$r(x) = \exp\left[\int_0^x (\alpha - \beta)\mathrm{d}x'\right]. \qquad (A.24)$$

Substituting the last result into (A.23) gives

$$\frac{\mathrm{d}q}{\mathrm{d}x} = -\alpha J \exp\left[-\int_0^x (\alpha - \beta)\mathrm{d}x'\right] \qquad (A.25)$$

with the solution

$$q(x) = -J\int_0^x \alpha \exp\left[-\int_0^{x'} (\alpha - \beta)\mathrm{d}x''\right]\mathrm{d}x' + C, \qquad (A.26)$$

where C is a constant to be determined by the boundary conditions. Multiplying the last equation by $r(x)$ we get

$$J_p(x) = -J \exp\left[\int_0^x (\alpha - \beta)\mathrm{d}x'\right] \int_0^x \alpha \exp\left[-\int_0^{x'} (\alpha - \beta)\mathrm{d}x''\right] \mathrm{d}x'$$

$$+ C \exp\left[\int_0^x (\alpha - \beta)\mathrm{d}x'\right]. \tag{A.27}$$

In the case the initial photon ionization occurs only in the p-region (see Fig. 5.15) so that the corresponding photon current of the minority carriers is $J_n(0)$. We apply the boundary condition $J_p(w) = 0$ and find

$$C = J \int_0^w \alpha \exp\left[-\int_0^x (\alpha - \beta)\mathrm{d}x'\right] \mathrm{d}x. \tag{A.28}$$

Further using (A.21) we also have the condition $J_p(0) = J - J_n(0)$ which is equal to C so that

$$J - J_n(0) = J \int_0^w \alpha \exp\left[-\int_0^x (\alpha - \beta)\mathrm{d}x'\right] \mathrm{d}x. \tag{A.29}$$

With the definition $M_n = J/J_n(0)$ we finally derive from (A.29)

$$M_n = \left\{1 - \int_0^w \alpha \exp\left[-\int_0^x (\alpha - \beta)\mathrm{d}x'\right] \mathrm{d}x\right\}^{-1}. \tag{A.30}$$

The derivation of M_p given by (5.92) is similar.

A.4 Power Flow of Standing Wave Modes

The incident radiation on a detector can be described by any set of orthogonal field functions that completely fill the space bounded by the planes containing the detector surface and the radiation source. Calculating the incident radiation power on a detector surface we imagine an arbitrary reflecting surface near the source at a distance d far enough from a detector to receive all wavefronts of the radiation at the detector parallel so that the field components on the detector are coherent. Standing waves of one spatial mode–called frequency modes–between this surface and the detector surface have fields for which the phase changes during a round-trip by a multiple of 2π or

$$2d = n\lambda = n\frac{c}{\nu}, \tag{A.31}$$

where n is an integer. The frequency difference between two adjacent modes is then $c/2d$. Having a bandwidth $\Delta\nu$ the number of standing modes is

$$N = \frac{2d\Delta\nu}{c}. \tag{A.32}$$

The total energy of these N modes is

$$E_{\mathrm{t}} = \frac{2d\Delta\nu}{c}\overline{E_{h\nu}} ,$$
(A.33)

where $\overline{E_{h\nu}}$ is the energy of a single frequency mode is given by (1.51). The power flow P is the total energy divided by the round-trip time $2d/c$ of this energy or

$$P = \frac{c}{2d}E_{\mathrm{t}} = \overline{E_{h\nu}}\Delta\nu .$$
(A.34)

It is seen that the power is proportional to the bandwidth and the arbitrarily chosen distance d is not relevant to the result.

References

1. Johnson, J.B.: Thermal agitation of electricity in conductors. Phys. Rev. **32**, 97 (1928)
2. Nyquist, H.: Thermal agitation of electric charge in conductors. Phys. Rev. **32**, 110 (1928)
3. van der Ziel, A.: Noise. Prentice-Hall, New York (1954)
4. van Vliet, K.M.: Noise in semiconductors and photoconductors. Proc. IRE **46**, 1004 (1958)
5. Langton, W.G.: A fast sensitive metal bolometer.J. Opt. Soc. Am. **36**, 355 (1946)
6. Billings, B.H., Hyde W.L, Barr, E.E.: An investigation of the properties of evaporated metal bolometers. J. Opt. Soc. Am. **37**, 123 (1947)
7. Mather, J.C.: Bolometer noise: nonequilibrium theory. Appl. Opt. **21**, 1125 (1982)
8. Wormser, E.M.: Properties of thermistor infrared detectors. J. Opt. Soc. Am. **43**, 15 (1953)
9. Low, F.J.: Low temperature germanium bolometer. J. Opt. Soc. Am. **51**, 1300 (1961)
10. Jones, R.C.: The general theory of bolometer performance. J. Opt. Soc. Am. **43**, 1 (1953)
11. Putley, E.H.: The pyroelectric detector. In: Willardson, R.K., Beer, A.C. (eds.) Semiconductors and Semimetals, vol. 5, pp. 259–285, Academic, New York (1970)
12. Schwarz, F., Poole, R.R.: Performance characteristics of a small TGS detector operated in the pyroelectric mode. Appl. Opt. **9**, 1940 (1970)
13. Baker, G., Charlton, D.E., Lock, P.J.: High performance pyroelectric detectors. Radio Electron. Eng. **42**, 260 (1972)
14. Sommer, A.H.: Photoelectric Materials. Wiley, New York (1968)
15. Scheer, J.J., van Laar, J.: GaAs-Cs, a new type of photo-emittor. Solid State Commun. **3**, 189 (1965)
16. Sonnenberg, H.: InAsP-CsO, a high efficient infrared photocathode. Appl. Phys. Lett. **16**, 245 (1970)
17. Csorba, Il.P.: Image Tubes. Howard W. Sams, Indianpolis, IN, USA (1985)
18. Putley, E.H.: Indium antimonide submillimeter photoconductive detectors. Appl. Opt. **4**, 649 (1965); also Putley, E.H.: InSb submillimeter photoconductive

devices. In Willardson, R.K., Beer, A.C. (Eds.) Semiconductors and Semimetals, vol. 12, pp. 143–168. Academic, New York (1977)

19. Cashman, R.J.: Film-type infrared photoconductors. Proc. IRE **47**, 1471 (1959)

20. Bartlett, B.E., Charlton, D.E., Dunn, W.E., Ellen, P.C., Jenner, M.D., Jervis, M.H.: Background limited photoconductive HgCdTe detectors for use in the 8–14 micron atmospheric window. Infrared Phys. **9**, 35 (1969)

21. Wang, S.Y., Bloom, D.M.: 100 GHz bandwidth planar GaAs Schottky photodiode. Electron. Lett. **41**, 211 (1982)

22. Rolls, W.H., Eddolls, D.V.: High detectivity $Pb_xSn_{1-x}Te$ photovoltaic diodes. Infrared Phys. **13**, 143 (1973)

23. McIntyre, R.J.: Multiplication noise in uniform avalanche diodes. IEEE Trans. Electron. Devices **ED-13**, 164 (1966)

24. Capasso, F., Tsang, W.T., Hutchinson, A.L., Williams, G.F.: Enhancement of electron impact ionization in a superlattice: A new avalanche photodiode with a large ionization rate ratio. Appl. Phys. Lett. **40**, 38 (1982)

25. Blauvelt, H., Margalit, S., Yariv, A.: Single-carrier-type dominated impact ionization in multilayer structures. Elec. Lett. **18**, 375 (1982)

26. Capasso, F., Tsang, W.T., Williams, G.F.: Staircase solid-state photomultipliers and avalanche photodiodes with enhanced ionization rates ratio. IEEE Trans. Electron Devices **ED-30**, 381 (1983)

27. Capasso, F.: Physics of avalanche photodiodes. In: Tsang, W.T (ed.) Semiconductors and Semimetals, vol.22D, pp 2–172. Academic, New York (1985)

28. Emmons, R.B.: Avalanche photodiode frequency response. J. Appl. Phys. **38**, 3705 (1967)

29. Hamstra, R.H., Wendland, P.: Noise and frequency response of silicon photodiode operational amplifier combination. Appl. Opt. **11**, 1539–1547 (1972)

30. Siegman, A.E.: The antenna properties of optical heterodyne receivers. Proc. IEEE. **54**, 1350–1356 (1966)

31. Ross, A.H.M.: Optical heterodyne mixing efficiency invariance. Proc. IEEE **58**, 1766–1767 (1970)

32. Kingston, R.H.: Detection of Optical and Infrared Radiation. Springer Series in Optical Sciences, vol.10 (1978)

33. Teich, M.C.: Three-frequency heterodyne system for aquisition and tracking of radar and communication signals. Appl. Phys. Lett. **15**, 420 (1969)

34. Abrams, R.L., White, R.C.: Three-freequency heterodyne detection of 10.6 μm laser signals. IEEE J. Quantum Electr. **QE-8**, 13 (1972)

35. Arechi, F.T., Gatti, E., Sona, A.: Measurement of low light intensities by synchronous single photon counting. Rev. Scient. Instrum. **37**, 942–945 (1966)

36. Bachman, R., Kirsch, H.D., Geballe, T.H.: Low temperature silicon thermometer and bolometer. Rev. Sci. Instr. **41** 547 (1970)

Index

amplifier noise, 28
  effective temperature, 30
  excess noise, 29
  noise figure, 29
autocorrelation
  signal recovery, 101
avalanche photodiodes, 86
  detectivity, 90
  frequency response, 92
  multiplication factor, 166
  multiplication process, 87
  noise, 89

background noise
  fluctuations, 44
bolometer, 36
  effective conductivity, 38
  metallic bolometer, 39
    Johnson noise, 39
  thermistor, 40
    Johnson noise, 42
    noise equivalent power, 43

correlation
  autocorrelation, 95
    spectral power density, 97
  cross correlation, 97
    spectral power density, 98
correlation computer, 119
cross correlation
  signal recovery, 99

dark current noise, 27
  Poisson statistics, 27

detection
  coherent versus incoherent, 129
detectivity, 21
  ideal detection, 24
    shot noise, 26
  specific detectivity, 21
dual signal beam, 138

effective temperature, 30
electron-hole pair, 11, 61, 64

fast detection, 153

g-factor, 12, 65
generation-recombination noise, 10, 13

heterodyne detection, 121
  beam profile, 124
  dual signal, 138
    lock-in amplifier, 145
    noise, 141
    noise equivalent power, 144
    waveform analyzing, 147
  incoherent radiation, 131
  intermediate frequency, 123
  lock-in amplification, 132
  lock-in amplifier
    noise, 133
    noise equivalent power, 136
  optical system, 127
  spectroscopy, 137
  thermal radiation, 130

Johnson noise, 2, 39, 43, 83, 154

laser radar, 149
linear detector system
　correlation, 105
lock-in amplifier, 111, 132, 145
　two phase, 115

micro pulse
　Fourier transform, 7, 8
　spectral power, 7
micropulse, 161

noise
　autocorrelation, 103
　Gaussian distribution, 154
　thermal radiation
　　amplitude and phase noise, 15
　　quantized noise, 15
　　spectral distribution, 17
　white noise, 7

operational amplifier, 108

P–N junction, 69
　current–voltage characteristic, 72
　saturation current, 75
　space charge region, 73, 77, 86
p–n junction, 70
　electron-hole pair, 75
　minority carriers, 72, 75
　photo current, 76
Peltier effect, 33
photocathode
　Fermi level, 51
photoconductors, 61
　detectivity, 67
　extrinsic
　　n-type, p-type, 11, 62
　frequency response, 69
　intrinsic, 11
　recombination, lifetime, 11
　responsivity, 66
　shot noise, 12
photodiode characteristic, 72
　current circuit, 82
　open circuit, 80
　reverse biased circuit, 83
photodiodes, 69
　current-voltage characteristics, 79
　detectivity, 85

efficiency, 77
　photon excitation, 75
photomultiplier, 56
　current gain, 56
　dark current, 56
　discriminator, 154
　noise currents, 58
　signal limitation, 59
　thermionic emission, 56
photon counting, 156
　discriminator, 157
　noise equivalent power, 159
　signal to noise ratio, 158
photons, 10
　fluctuations, 10
PIN diodes, 78, 86
Planck's law, 3
Poisson distribution, 22, 163
pulse train averager, 116
pyroelectric detector, 44, 131
　noise equivalent power, 48
　pyroelectric coefficient, 45
　responsivity, 45, 46

Rayleigh distribution, 16
　Rayleigh noise, 17

semiconductors
　intrinsic, extrinsic, 63
sensitivity, 28
shot noise, 5
　autocorrelation, 104
　current fluctuations, 5
　spectral distribution, 6
signal averager, 115
signal processing, 107
signal-noise relations
　amplifier limitation, 27
　background limitation, 22
　dark current limitation, 27
　signal limitation, 22
single frequency mode, 14
space communication, 148
spectroscopy, 137
standing wave modes, 167
statistical thermodynamics, 17
statistics, 162
　binomial distribution, 162
　Gaussian distribution, 164

photoelectrons, 165
Poisson distribution, 163
Stefan-Boltzmann law, 14

temperature fluctuations, 18
  absorption, emission, 20
  power spectrum, 19
thermal detectors, 31
thermal noise, 2
  capacitor, 4
  resistor, 3
thermal radiation, 13
  spatial mode, 15
  standing wave modes, 167
thermistor, 40
  electrothermal feedback, 43
thermocouple, 31, 36
  Johnson noise, 34

responsivity, 34
  thermoelectric power, 32
thermopile, 36
transmission line, 2
transmitting photograph, 149

vacuum photodetectors, 51
  electron affinity, 51
  photocathode, 51
vacuum photodiode, 52
  dark current, 54
  frequency response, 54
  noise spectrum, 9
  transit time, 53

waveform analyzer, 118, 147
Wiener-Khintchine theorem, 97

Springer Series in
ADVANCED MICROELECTRONICS

1 **Cellular Neural Networks**
Chaos, Complexity
and VLSI Processing
By G. Manganaro, P. Arena,
and L. Fortuna

2 **Technology of Integrated Circuits**
By D. Widmann, H. Mader,
and H. Friedrich

3 **Ferroelectric Memories**
By J.F. Scott

4 **Microwave Resonators and Filters
for Wireless Communication**
Theory, Design and Application
By M. Makimoto and S. Yamashita

5 **VLSI Memory Chip Design**
By K. Itoh

6 **Smart Power ICs**
Technologies and Applications
Ed. by B. Murari, R. Bertotti,
and G.A. Vignola

7 **Noise in Semiconductor Devices**
Modeling and Simulation
By F. Bonani and G. Ghione

8 **Logic Synthesis for Asynchronous
Controllers and Interfaces**
By J. Cortadella, M. Kishinevsky,
A. Kondratyev, L. Lavagno,
and A. Yakovlev

9 **Low Dielectric Constant Materials
for IC Applications**
Editors: P.S. Ho, J. Leu, W.W. Lee

10 **Lock-in Thermography**
Basics and Use
for Functional Diagnostics
of Electronic Components
By O. Breitenstein
and M. Langenkamp

11 **High-Frequency Bipolar Transistors**
Physics, Modelling, Applications
By M. Reisch

12 **Current Sense Amplifiers**
for Embedded SRAM
in High-Performance
System-on-a-Chip Designs
By B. Wicht

13 **Silicon Optoelectronic
Integrated Circuits**
By H. Zimmermann

14 **Integrated CMOS Circuits
for Optical Communications**
By M. Ingels and M. Steyaert

15 **Gettering Defects
in Semiconductors**
By V.A. Perevostchikov
and V.D. Skoupov

16 **High Dielectric Constant Materials**
VLSI MOSFET Applications
Editors: H.R. Huff and D.C. Gilmer

17 **System-level Test and Validation
of Hardware/Software Systems**
By M. Sonza Reorda, Z. Peng,
and M. Violante